Management and Industrial Engineering

Series editor

J. Paulo Davim, Aveiro, Portugal

For further volumes:
http://www.springer.com/series/11690

Carolina Machado · J. Paulo Davim
Editors

Work Organization and Human Resource Management

Springer

Editors
Carolina Machado
Department of Management
School of Economics and Management
University of Minho
Braga
Portugal

J. Paulo Davim
Department of Mechanical Engineering
University of Aveiro
Aveiro
Portugal

ISBN 978-3-319-35301-2 ISBN 978-3-319-06376-8 (eBook)
DOI 10.1007/978-3-319-06376-8
Springer Cham Heidelberg New York Dordrecht London

Printed on acid-free paper

Springer is part of Springer Science+Business Media (www.springer.com)

Preface

Work Organization and Human Resource Management covers issues related to the new advances in work organization and human resource management that result from the continuous and highly complex process of change that today's organizations are facing. In the present, as a consequence of social, technological, political and economic changes, the field of work and organizations management becomes highly complex, claiming management for more effective strategies. Conscious of this reality this book looks to provide support to academics and researchers, as well as those operating in the management field, to deal with policies and strategies related to work issues and human resource management.

Effective work practices and good employee relations are a real necessity of today's organizations, as they can help to reduce absenteeism, turnover, organizational costs, conducting to high levels of commitment, effectiveness, performance as well as productivity. Addressing these questions, this book focuses on the implications of those changes in work organization and human resource management.

Drawing on the latest developments, ideas, research and the best practice, the book intends to examine the new advances in work organization and human resource management resulting from the recent changes that are taking place and how they affect the management as well as the commitment and motivation of these organizations' workers.

This book covers *Work Organization and Human Resource Management* in 11 chapters. Chapter 1 discusses "The Role of Responsible HRM Practices and a Culture-Related Capability on the CSR-Performance Association: A Small Firm Perspective". Chapter 2 contains information about "Managing High Performance Work Systems and Organizational Performance". Chapter 3 covers "The Role of the Line Manager in HRM-Performance Research". Chapter 4 describes "The Impact of Strategic Human Resource Management Practices on Australian Dairy Farm Performance". Subsequently, Chapter 5 covers "Exploring Change in Small Firms' HRM Practices". Chapter 6 contains information on "Effective Human Resource Practices in Family Businesses". Chapter 7 describes "The Gender of the Entrepreneur and the Politics of Entrepreneurship for Women: An Exploratory Study". Chapter 8 focuses on the "Influence of Organizational Flexibility in High Performance Work Practices". Chapter 9 discusses "Gaining or Losing? Projective Identification, Professional Identities and New Public Management". Chapter 10 focuses "Researching HRM to Enhance Understanding: The Neglected Role of Ontology".

In Chapter 11, "Examining the Motivation to Work Within a Group of Gardeners with Mental Disorders" is presented.

Finally, it is important to say that this book is designed to increase the knowledge and effectiveness of all those involved in human resource management and working conditions in all kinds of organizations and activity sectors.

Whether we are a human resource manager, a manager, an engineer, a policy maker, a strategist, a practitioner, an academic or a researcher, we need to know what is happening, on both national and international environments, to be able to understand and develop effective responses to meet all these new demands and challenges. This is why, today, the interest in this subject is evident for many types of organizations, namely, important institutes and universities all over the world.

The Editors acknowledge their gratitude to Springer for this opportunity and for their professional support. Finally, we would like to thank all the chapter authors for their interest and availability to work on this project.

Braga, Portugal Carolina Machado
Aveiro, Portugal J. Paulo Davim

Contents

Contributors

Aykut Berber Department of Management and Organization, School of Business, Istanbul University, Istanbul, Avcilar, Turkey

Kenneth Cafferkey Graduate School of Business, Universiti Tun Abdul Razak, Kuala Lumpur, Malaysia

João Fontes da Costa Estgoh, Polytechnic Institute of Coimbra, Oliveira do Hospital, Portugal

Teresa Carla Trigo de Oliveira Faculty of Economics, University of Coimbra, Coimbra, Portugal

Emília Fernandes Department of Management, School of Economics and Management, University of Minho, Braga, Portugal

Santiago Gutiérrez-Broncano Faculty of Social Sciences, Department of Business Administration, University of Castilla-La Mancha, Talavera de la Reina, Toledo, Spain

Brian Harney Dublin City University Business School, Dublin City University, Dublin, Ireland

Pedro Jiménez-Estévez Faculty of Social Sciences, Department of Business Administration, University of Castilla-La Mancha, Talavera de la Reina, Toledo, Spain

Regina Leite Department of Management, School of Economics and Management, University of Minho, Braga, Portugal

Carolina Feliciana Machado Department of Management, School of Economics and Management, University of Minho, Braga, Portugal

O. Mallett Durham University Business School, Durham University, Durham, UK

Wayne O'Donohue Department of Employment Relations and Human Resources, Griffith University, Gold Coast, Australia

Ester Machado Pereira School of Economics and Management, University of Minho, Braga, Portugal

Mercedes Rubio-Andrés Department of Business, School of Business and Economics, San Pablo CEU University, Madrid, Spain

D. Spicer School of Management, Bradford University, Bradford, UK

Nuttaneeya (Ann) Torugsa Australian Innovation Research Centre, University of Tasmania, Hobart, Australia

Aman Ullah Department of Livestock Economics and Business, University of Veterinary and Animal Sciences, Lahore, Pakistan

R. Wapshott Management School, Sheffield University, Sheffield, UK

Murat Yaslioglu Department of Management and Organization, School of Business, Istanbul University, Istanbul, Avcilar, Turkey

Connie Zheng Faculty of Business and Law, Deakin Graduate School of Business, Deakin University, Burwood, VIC, Australia

The Role of Responsible HRM Practices and a Culture-Related Capability on the CSR-Performance Association: A Small Firm Perspective

Wayne O'Donohue and Nuttaneeya (Ann) Torugsa

Abstract This chapter focuses on the role that responsible human resource management practices (HRM) play in relation to proactive corporate social responsibility (CSR) and its impact on firm performance, using evidence from small firms in the Australian manufacturing industry. The empirical evidence reveals the critical moderating contribution of responsible HRM practices in the implementation of proactive CSR and enhancement of financial performance in small firms. The study findings indicate the need for managers of small firms, who wish to boost financial competitiveness through proactive CSR, to adopt responsible HRM practices that avoid a directive approach and which foster employee participation and engagement. It is suggested that this can be done by leveraging the relative informality of work organisation and familial ties that help engender trust and reciprocity in small firms.

1 Introduction

In today's business climate, a singular emphasis by firms on management practices that maximise profitability, whilst ignoring their wider negative impacts over the long-term, is considered generally no longer acceptable. Increasingly, firms regardless of size are being expected to recognise their social responsibility, take the initiative, and contribute towards sustainability [1, 2]. This implies the necessity for firms in today's highly competitive business environment to adopt management practices

W. O'Donohue (✉)
Department of Employment Relations and Human Resources,
Griffith University, Gold Coast, Australia
e-mail: w.odonohue@griffith.edu.au

N. Torugsa
Australian Innovation Research Centre, University of Tasmania, Hobart, Australia

C. Machado and J. P. Davim (eds.), *Work Organization and Human Resource Management*, Management and Industrial Engineering,
DOI: 10.1007/978-3-319-06376-8_1, © Springer International Publishing Switzerland 2014

that deliver not only superior financial performance but also create sustainable outcomes of value that meet societal expectations [3]. The role played by a firm's human resource management (HRM) practices in relation to sustainability and social responsibility, and its impact on firm performance has received broad research attention in the literature [4–6]. If HRM in general terms is considered to be the 'productive use of people in achieving the organization's strategic objectives and the satisfaction of individual employee needs' [7], there is still much to learn about the process through which a firm's HRM practices interact with other organisational practices to deliver performance outcomes that result in competitive advantage and superior financial returns for the firm [6].

This paper addresses that gap in understanding by examining the moderating contribution of a firm's HRM—considered in terms of responsible practices and a culture-related capability—to the implementation of corporate social responsibility (CSR) for generating an improvement in business financial performance, using evidence from small firms in the machinery and equipment sector of the Australian manufacturing industry. For the purpose of this study, the number of employees is used to determine firm size [8]; hence in line with previous research [9, 10], a small firm is defined as having an upper limit of less than 100 employees. Firms of this size are of interest because the role of HRM as a moderating influence on the CSR-performance association is under-researched when compared to medium and large sized firms [11, 12].

2 Role of HRM in the CSR-Performance Association

Many HRM scholars contend that effective management of a firm's human resource capability is fundamental to the achievement of sustainable and socially responsible outcomes; indeed, CSR sits at the very heart of their conception of HRM as an essential organisational practice. Advocates of this view contend that the three principles of sustainability form the nexus between high quality management of a firm's human resources and the creation of high quality transparent and ethical relationships both inside and external to the firm. This nexus is held to operate in all firms regardless of size [6, 13–16]. Before considering the role of HRM let us make clear some of the definitional boundaries within which the discussion will occur.

2.1 Corporate Social Responsibility and Financial Performance

Whilst definitions of corporate social responsibility and sustainability might originally have differed, there has in recent years been a convergence of meaning in the literature to the point where the terms may be used interchangeably [2, 17]. Given this convergence, in this paper we have chosen to use the term 'corporate social responsibility' (CSR) defined as responsible management practices that support the three principles of sustainability: economic growth and prosperity, social cohesion and equity, and environmental integrity and protection [18]. Under this definition, if firms are to meet their societal responsibilities and deliver sustainability, a responsible management approach is required; one which acknowledges the need for profitability, but at the same time takes an integrated, holistic perspective that emphasises outcomes and practices that support the three principles of sustainability equally and satisfies a broad range of stakeholders [2, 18]. In a competitive business environment simply maintaining a reactive stance to CSR-related legal requirements, with management practices that aim solely at achieving regulatory compliance and maintaining legitimacy, arguably will not suffice; instead, adoption of a range of responsible proactive management practices that support the economic, social and environmental principles of sustainability has been advocated as being a significant priority for firms seeking a competitive advantage and enhanced performance [19–24]. It is this 'proactive' form of CSR (i.e. voluntary adoption of responsible management practices that go beyond regulatory and legitimacy requirements, in order to enhance firm performance as well as contribute broadly and positively to society) that is of interest in this study [25, 26].

Responsible management embracing the principle of economic sustainability connects most directly to a firm's profitability [18]. Characteristically, such management practices eschew a narrow focus on managing short-term financial performance (profit maximising) and emphasize active management of long-term economic performance issues that are important, not just for the firm but for all stakeholders in its host community. In terms of purpose, economic-related CSR aims, for example, at encouraging innovation, efficiency, and the effective creation and sustainable management of a firm's economic capital, thereby fostering longevity in profitability and growth. Firms engaging in proactive management of economic-related CSR issues deliberately pay attention to a range of matters such as innovative capability, product quality, fair pricing, and supply chain management from sourcing to final payment [27–29].

Responsible management supporting the environmental principle of sustainability focuses generally on eco-efficiency and pollution prevention, and aims at minimising the firm's ecological footprint at all points in its product life cycle [30]. Such activity is often characterized by adoption of internationally standard environmental management systems (or a total quality environmental management approach) that ensure a firm's environmental impacts are monitored and managed systematically rather than on an ad hoc basis [31]. Systematic environment-related CSR helps build a firm's credibility among its external stakeholders, as well as ensure that the principle of environmental sustainability is institutionalized among employees [30].

Turning to the third CSR principle of social sustainability, responsible management in this dimension can be categorised as having either an external focus on the broader community, or an internal focus on the workplace and employees [18]. Accordingly, social-related CSR that focuses on the link between the firm and its external community, such as philanthropic and corporate citizenship initiatives, is informed by the general value systems, norms and pressures which dominate social community networks in which firms operate [3, 32, 33]. At the same time, social-related CSR reflects a focus in the workplace on the health, safety, and personal and professional well-being of employees; it is the mechanism through which internal workplace human resource issues, such as work/life balance, employee participation in decision making process, and diversity and equal opportunity, are managed [18, 32, 34].

The amount of empirical research examining the effectiveness of the proactive form of CSR as a business management practice that delivers superior financial returns is limited but growing. To date, whilst the research overall has demonstrated that CSR can impact positively on financial performance, much of the research has been conducted with a specific and narrow focus on environmentally related CSR and mainly in large firms [35–37]. This focus is similar in the much less studied small firm context, where quantitative research has found a positive link between financial performance and proactive CSR (in its environmental dimension). In addition, a qualitative study of social- and environment-related CSR has reported that small firms can derive not only quantifiable benefits (e.g. cost savings and increased profitability) but also, and in the main, a range of intangible benefits (e.g. improved image and reputation, and enhanced trust) [34]. These research findings, in line with other studies of firm performance in the literature, show that embedding the economic, social and environmental principles of sustainability in proactive management practices, can accrue non-material benefits to the firm, such as social approval and reputation building, improved customer relationships, and increased access to resources, all of which together positively influence firm performance [2, 20, 21, 27, 38].

Most recently and of specific interest here, Torugsa and colleagues have presented evidence indicating proactive CSR has a positive effect on financial performance in small and medium enterprises (SMEs); they also found that the influence of each of the three CSR dimensions is positively moderated by an interactive process between the three dimensions of CSR [26]. Based on their results, they concluded that SMEs should pay primary attention to economic-related proactive CSR, while simultaneously and selectively focusing on supportive elements of social- and environment-related proactive CSR, if sustainable long-term financial performance is to be achieved [26]. In the light of these research findings, this study focuses on the role played by a firm's HRM practices—that is one of the two elements in social-related CSR outlined above—in the delivery of CSR performance outcomes that enhance financial returns. Drawing on the HRM and resource-based view literature, we will now discuss and develop some hypotheses regarding the links between responsible HRM practices and the effective adoption of proactive CSR across its three dimensions.

2.2 Moderating Role of HRM: Responsible Practice and Capability Perspectives

Basically, the centrality of human resources in CSR is argued on three grounds. Firstly, in terms of the sustainability principle of economic growth and prosperity, ongoing innovation is seen as the major factor that underpins a firm's performance.

In particular, evidence for the fundamental role and contribution of a firm's human resource capability in the creation and support of innovation as a competitive advantage has been well documented in the literature [3, 4, 16, 39, 40]. Similarly, there is much recent empirical research that supports the crucial role played by specific HRM practices (e.g. training and development, talent management, organisational culture and development) in stimulating innovation in support of sustainable outcomes linked to financial performance [14, 41, 42].

Secondly, there is research highlighting the importance of human resources in the achievement of environmental sustainability and a firm's performance [14, 43, 44]. Human resource attributes, such as employee attitudes, environmental values, technical and managerial skills have been shown to be fundamental to the development and implementation of proactive environmental CSR [31, 45–47]. Across this research, there is general agreement that high quality HRM practices can motivate and assist employees to adopt and enact the principle of environmental sustainability in their work. The means to this end include managing organisational culture and development so as to integrate environmental sustainability into the firm's values and norms, and demonstration of a credible commitment by the firm to continuous improvement of employee environmental management skills and knowledge.

Thirdly, and perhaps most importantly, the case for the centrality of human resources in CSR is grounded on the significant commonality in issues that responsible HRM practices and social CSR practices in the workplace in particular seek to address. For example, scholars contend that a key human resources concern in today's business environment is internal diversity management; a concern they also identify as being *inter alia* central to sustainability and a firm's performance [14]. Research also maintains that regeneration of value and the renewal of wealth underpin sustainability; accordingly, issues such as equity, justice, respect, employee consultation, work-life balance, development, and employee well-being, all of which are grist for the HRM mill, must be addressed in the pursuit of sustainability and associated improvement in a firm's performance [13, 14, 48]. Building on the HRM literature [49, 50], Shen and Zhu [51] describe the CSR-related concept of socially responsible human resource management practices that embraces workplace health and safety, equity, work-life balance, employee development, and employee involvement and participation. Finally, Gond and colleagues [52] offer a model (see Fig. 1), which clarifies the relationship between CSR and responsible HRM practices.

Figure 1 indicates there are activities which form a focus solely for HRM or CSR, and these activities are differentiated from other activities that involve mutually supportive actions by one domain to the other. At the heart of the model are those activities common to both domains [52]. Given the consensus in both the CSR and HRM literature, we consider 'social CSR in the workplace' and 'responsible HRM practices' to be in effect one and the same thing; this implies responsible HRM practices are a key moderating factor in the relationship between CSR in all three of its dimensions and a firm's performance generally, and financial performance specifically.

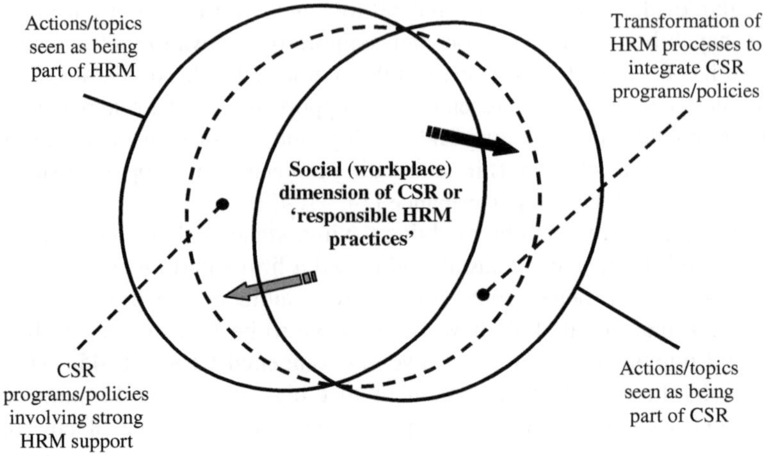

Fig. 1 Responsible HRM and social (workplace) CSR dimension (adapted from [52])

In line with the above discussion of the contribution of CSR to financial performance, and the distinction we draw between external social CSR in the community and internal CSR in the workplace (also called responsible HRM practices), we propose the following hypotheses for testing in relation to small firms:

Hypothesis 1: Economic and environmental CSR together with social CSR in the community (EcEnvSoC) is positively associated with financial performance

Hypothesis 2: Responsible HRM practices (RHRM) positively moderate the association between EcEnvSoC and financial performance in such a way that the positive association is significantly stronger at high levels of RHRM

In line with the resource-based view of the firm, it is not only a firm's resources (e.g. financial, human and physical) but also its capabilities (i.e. the integrated application of its resources) that drive firm performance [53, 54]. While a wide variety of specific capabilities that support proactive CSR and financial performance have been discussed in the research literature (such as 'shared vision and employee involvement' [2, 55, 56]; 'stakeholder management' [57]; 'innovation' [58, 59]; 'strategic proactivity' [31]; and 'higher-order learning' [60]), much of the discussion has focused predominantly on capabilities in relation to only one dimension of CSR—the environmental dimension—and its management in large firms. In contrast, in their study of Australian manufacturing SMEs, Torugsa and colleague [25, 26] examined three specific capabilities ('shared vision', 'strategic proactivity' and 'stakeholder management') and the association between those capabilities, the adoption of CSR in its three dimensions, and financial performance. They found that these organisational capabilities underpinned the adoption and implementation of proactive CSR as a business strategy that linked to improved financial performance in SMEs.

Given our focus on responsible HRM practices in small firms, we restrict our discussion specifically to those two capabilities considered by [25, 26] which have been identified most closely in the literature with HRM [3, 4, 61]. The first of the capabilities—'*Shared Vision*'—exists when the members of a firm have a shared feeling that the firm's objectives are important and that all of its members may contribute to defining them. Such a capability is held to be fundamental to the different parts of a firm being able to bond together and integrate or combine resources effectively [62]. Creating a shared vision that supports proactive CSR entails recognition by a firm of the essential role of responsible HRM practices, and the need for a deep commitment to empowering social processes that generate consensus and significant employee involvement [56, 63]. Indeed, many scholars argue that the creation of a shared vision must be a primary goal of responsible HRM practices [64, 65].

The second capability—*'Strategic Proactivity'*—is realised when a firm, rather than merely reacting to changes in its business environment, uses responsible management practices to shape the environment to its own advantage, and identify and capitalize on new emerging opportunities [1, 31, 59, 66–68]. As with the case of shared vision, a strategic proactivity capability is closely linked to a firm's organisational culture which is derived in large part from a purposeful strategic approach to managing its human resources [61]; strategic proactivity requires a commitment to responsible HRM practices that empower, develop and involve employees, as well as encourage and guide employees to think and act proactively in line with the firm's CSR business strategy [31, 68]. Given the complementarity of these two capabilities and their importance as a focus for responsible HRM practices, for the purposes of this study we consider them as forming a single organisational culture-related capability which we will call a *'Shared Strategic Vision'* capability.

In accord with other scholars [13, 14, 69], we suggest the alignment of a shared strategic vision capability and responsible HRM practices with the economic, social and environmental principles of sustainability is 'natural', and creates the potential for responsible HRM practices to be the 'tipping factor' that leads to the successful execution of CSR as a business strategy producing superior financial performance. Consequently, we propose the following hypothesis for testing:

> *Hypothesis 3:* The association proposed in Hypothesis 2 is positively moderated by a firm's capability of shared strategic vision (SSV) in such a way that the positive interaction effect of EcEnvSoC and RHRM on financial performance is greater at high levels of SSV.

3 Method

The sample population comprised 1,278 Australian small firms, in the machinery and equipment manufacturing sector, drawn from a commercial database [70]. Small firms in this sector were attractive as a focus of study for three main reasons.

First, while the role of responsible HRM practices in the implementation of CSR has been well researched in larger enterprises, smaller firms have received far less research attention [12]. Second, in theory the more flexible and less bureaucratic organisational structures of small firms might make such firms more proactive in their engagement in CSR [68]. Finally, the significant socio-environmental impacts of firms in this sector (due to the nature of processes used to transform raw materials into finished manufacturing products) make them more likely to experience pressure to engage 'explicitly' in CSR activities.

This study employed a survey-based method for testing the research hypotheses. A survey questionnaire was developed based on the extant literature and existing published questionnaire items, and was pretested with owner-managers of three small firms to ensure content validity. As in small firms decision-making is often highly centralized, surveying a single experienced well-qualified informant is argued to better capture a firm's approach than surveying several informants [71, 72]. Hence, a single informant in each firm (either the business owner and/or CEO, managing director) was used in this study. The survey was administered by mail in April, 2009 (Time 1) and November, 2009 (Time 2), providing a six-month time lag between the measurements of predictor/moderators (EcEnvSoC, RHRM and SSV) and the dependent variable (financial performance). Such data collection allowed us to perform a bivariate correlation analysis between the same variable at the two different time points, thereby enabling reliability of the data to be evaluated. The six month delay in the data collection process also increased the likelihood that sufficient time would have elapsed to require informants to provide updated responses rather than simply repeat previous Time 1 responses.

The survey received 183 responses (from a possible 1,278 responses), representing a 14.3 % response rate. After responses with missing data were eliminated, a total of 158 firms remained for analysis. Of the firms that did respond, the majority employed between 10 and 49 employees (75.3 %) and had annual turnover between $AU1 and $AU10 million (66.5 %). Business owners (67.1 %) and respondents who had more than 5 year experience (74.7 %) in their firms were the majority of respondents.

Based on previous literature [73, 74], it was recognised that data reported by single informants may be subject to common method bias. To test whether this was a problem, Harman's single-factor test was performed on the final sample through an exploratory factor analysis [75]. Results of the analysis showed no single general factor as accounting for most of the covariance in the variables, thus suggesting the interpretation of our results would not be confounded by common method bias. Moreover, an advantage of the moderated hierarchical regression analysis method employed in this study, was that common method effects are dealt with (along with main effects) before inspecting any interaction terms [76]. We also assessed non-response bias through the time-trend extrapolation procedure [77]; and no significant difference was detected between early and late respondents in terms of their demographic characteristics.

3.1 Measures

3.1.1 Corporate Social Responsibility (EcEnvSoc) and RHRM

In the absence of any publicly available data on CSR in small Australian manufacturing firms, respondents' perceptions of their firms were measured using an instrument comprising 27 survey items based on: the extant literature [18, 27, 28, 34, 59, 78]; and feedback received from pre-test participants. Respondents were asked to indicate the extent to which their firms voluntarily engaged in CSR activity compared to similar firms in their industry sector, using a five-point scale (1 = 'not addressed issue at all' to 5 = 'we are leaders on this issue'). A bivariate correlation analysis revealed the high correlation between Time 1 and Time 2 CSR data ($r = 0.95$), thus confirming our confidence in the reliability of the CSR scale.

An exploratory maximum likelihood factor analysis with varimax rotation was conducted and seven factors with eigenvalues greater than 1 emerged. The factors were related to different groups of CSR practices, and each of the 27 measured variables had a significant factor loading on one of these factors. Cronbach's alpha (α) ranged from 0.716 to 0.873 for each factor. A confirmatory factor analysis of items pertaining to the seven underlying identified factors (see Table 1) showed a good fit to the data ($\chi^2 = 358.67$; df = 274; RMSEA = 0.039; CFI = 0.99; IFI = 0.99; and NNFI = 0.99) and convergent validity (all standardized factor loadings above the value of 0.50).

Factors 1–6 (as shown in Table 1) were considered indicators of a single factor, which we labelled '*economic and environmental CSR, together with social CSR in the community*' (EcEnvSoC). A second-order confirmatory factor analysis showed that the model fitted the data well ($\chi^2 = 16.06$; df = 11; RMSEA = 0.057; CFI = 0.99; IFI = 0.99; and NNFI = 0.99), and all six factors were significantly ($p < 0.01$) related to EcEnvSoC with standardized loadings ranging from 0.66 to 0.92. The final measure of EcEnvSoC was a weighted average of the six factors using the standardized loadings obtained from the second-order confirmatory factor analysis, and a high score was indicative of a high degree of EcEnvSoC. Factor 7 (as shown in Table 1) was considered an indicator of '*responsible HRM practices*' (RHRM) given the internal focus of the constituent items. The final measure of this variable was a weighted average of five underlying identified factors (SOC1-5) using the standardized loadings obtained from the confirmatory factor analysis, and a high score was indicative of a high degree of RHRM.

3.1.2 Shared Strategic Vision (SSV)

A culture-related organisational capability was measured using a compilation of the three shared vision capability items from [78], and the three strategic proactivity capability items from [31]. All six items were presented as statements related to the

Table 1 Results of confirmatory factor analysis of CSR items

Factor	Measured variable	Factor loadings (λ)*	Reliability (α)
Economic and environmental CSR together with social CSR in the community (EcEnvSoC)			
Factor 1	ECON2: adopt a long-term perspective in decision-making in order to guarantee sufficient cashflow and produce a persistent superior return to shareholders/owners	0.638	0.758
	ECON6: use of certification on quality aspects e.g. ISO 9000	0.631	
	ECON7: responsible supply chain management, from sourcing to final payment e.g. meeting payment schedules	0.706	
	ECON8: create spin-off technologies that can be profitably applied to other areas of the business	0.549	
Factor 2	ECON1: work with government officials to protect the firm's interest	0.680	0.716
	ECON3: reduce costs of inputs for the same level of outputs	0.504	
	ECON4: differentiate product/process by marketing of the social and environmental performance of the product/process	0.709	
	ECON5: sell waste products for revenue	0.518	
Factor 3	ENV1: periodic natural environment audits	0.798	0.873
	ENV2: purchasing criteria including ecological requirement	0.699	
	ENV3: environmental training for employees	0.575	
Factor 4	ENV4: filters and controls on emissions and discharges	0.629	0.840
	ENV6: program of waste recycling/reuse	0.741	
	ENV8: reduction/replacement of hazardous chemicals or materials e.g. substituting hazardous chemicals with less hazardous alternatives	0.562	
	ENV9: systematically separate dangerous wastes	0.651	
Factor 5	ENV5: program for water recycling	0.508	0.795
	ENV7: increase energy efficiency	0.619	
	ENV10: use of certifications on environmental aspects e.g. ISO 14000	0.546	
	ENV11: design products and manufacturing processes to minimise the ecological footprint along the entire product life cycle	0.653	

(continued)

Table 1 (continued)

Factor	Measured variable	Factor loadings (λ)*	Reliability (α)
Economic and environmental CSR together with social CSR in the community (EcEnvSoC)			
Factor 6	SOC6: engage in philanthropic activities e.g. charitable donation	0.737	0.832
	SOC7: sponsorship of local community initiatives	0.832	
	SOC8: consider interests of stakeholders in investment decisions by creating a formal social dialogue	0.501	
Responsible HRM practices (RHRM)			
Factor 7	SOC1: employee participation in decision-making process	0.679	0.790
	SOC2: creation of good work-life balance and family friendly employment	0.701	
	SOC3: investor in people e.g. training and employee development	0.709	
	SOC4: equal opportunities in workplace e.g. employing disabled people, and/or promoting women to senior management positions	0.719	
	SOC5: improve employee health and safety	0.710	

Note Within the EcEnvSoc dimension, factors 1–2 represent economic CSR; factors 3–5 represent environmental CSR; and factor 6 represents social CSR in the community. All factor loadings are significant at $p < 0.001$

Table 2 Shared strategic vision

	Item	Factor loading (λ)	Reliability (α)
Shared vision	SHAR1: the objectives of this firm are very well-known to everybody working here	0.611	0.703
	SHAR2: everybody working in this firm influences the way to work and the objectives of the firm	0.811	
	SHAR3: everybody in this firm freely contributes his/her points of view about how to run it smoothly	0.579	
Strategic proactivity	STRA1: our products are many and very different. We are always looking for new opportunities i.e. in very different areas in the manufacturing industry	0.645	0.720
	STRA2: the main technology focus of this firm is on having leading flexible and innovative technologies	0.709	
	STRA3: our planning systems are very open and flexible to allow us to seize new opportunities	0.693	

firm, against each of which respondents were asked to rate their level of agreement on a six-point scale (1 = 'strongly disagree' to 6 = 'strongly agree') (see Table 2). The correlations obtained for each set of three items (shared vision—$r = 0.89$; strategic proactivity—$r = 0.89$) between Time 1 and Time 2 were high thus confirming our confidence in the reliability of each scale.

As outlined earlier, for the purposes of this study, a shared vision capability (Cronbach's $\alpha = 0.703$) and strategic proactivity capability (Cronbach's $\alpha = 0.720$) were considered indicators of a single factor, which we labelled '*shared strategic vision*' (SSV). A second-order confirmatory factor analysis revealed that the model had a good fit to the data ($\chi^2 = 12.04$; df = 7; RMSEA = 0.053; CFI = 0.99; IFI = 0.98; and NNFI = 0.98), and these two factors were significantly related to SSV (standardized loadings of shared vision = 0.69 and of strategic proactivity = 0.76, $p < 0.01$). The final measure of SSV was the average of the two factors using the standardized loadings obtained from the second-order confirmatory factor analysis, and a high score was indicative of a high degree of SSV.

3.1.3 Financial Performance

Our pre-test discussions with small business' owner managers revealed that respondents would be unlikely to provide specific objective data on their firm's financial performance due to its commercial-in-confidence in nature. In accordance with the literature that shows a high correlation and concurrent validity between objective and subjective data on performance, implying that both are valid when

Table 3 Financial performance

Item	Factor loading (λ)	Reliability (α)
FINA1: return on assets (earnings generated from invested assets)	0.888	0.912
FINA2: net profits to sales	0.974	
FINA3: liquidity (cash flows and/or access to funds)	0.793	

calculating a firm's financial performance [79, 80], we thus collected the respondents' perceptions on three financial performance items—return on assets, net profits to sales and liquidity (Cronbach's $\alpha = 0.912$). These items were drawn from prior research [79, 81, 82]. Respondents were asked to rate their firm's financial performance, over the preceding six-month period compared to similar firms in their industry sector, using a five-point scale (1 = 'much worse' to 5 = 'much better') (see Table 3). The factor score of financial performance was a weighted average (using the standardized loading) of these three items, with a high score being indicative of a high level of financial performance. In the absence of publicly available objective data, the high correlation obtained for financial performance between Time 1 and Time 2 surveys ($r = 0.86$) confirmed our confidence in the reliability of the financial performance scale.

3.1.4 Control Variables

Although we restricted our analysis to small firms, the breadth of this category suggested the need to use firm size to control for potential differences that may affect financial performance [59, 78]. The size of a firm (measured by the number of employees employed on a regular basis) was thus controlled in this study. Furthermore, as our study was conducted during the global financial crisis (GFC), the potentially negative impact on firm performance of this external influence was treated as another control variable. This control variable was measured in terms of the extent to which general economic conditions had negatively impacted in the previous six-month period in relation to return on assets, net profits to sales and liquidity—using a five-point scale (1 = 'no impact at all' to 5 = 'very high impact').

4 Analysis and Results

4.1 Descriptive Statistics

Analysis of the survey responses relating to CSR performance overall (EcEnvSoC plus RHRM) shows 39.5 % of the sample firms reported the highest performance levels (i.e. ahead of, or leading, similar firms in their industry), with 31.5 % reporting

Table 4 Means, standard deviations and correlations (N = 158)

	Mean	SD	1	2	3	4	5
1. EcEnvSoC	2.87	0.74					
2. RHRM	3.56	0.68	0.56***				
3. SSV	3.64	0.57	0.40***	0.44***			
4. Financial performance	3.38	0.95	0.48***	0.26**	0.25**		
5. Firm size	2.57	0.86	0.21**	0.14	-0.02	0.27**	
6. Perceived GFC	2.75	0.86	-0.10	-0.09	-0.04	-0.49***	-0.06

Note $^{**}p < 0.01$; $^{***}p < 0.001$

the lowest performance levels (i.e. not addressed at all or behind similar firms). Looking at each specific dimension of CSR, the highest number of firms reporting they were ahead of, or leading, similar firms was 58.7 % for RHRM, compared with 48.2 % for economic CSR, 27.9 % for environmental CSR, and 25.3 % for social CSR in the community. In terms of leadership in each dimension, the reported emphasis differed, with economic CSR ranking highest at 14.7 %, followed by RHRM (11 %), and environmental and community-related social CSR (4.9 % and 4.6 % respectively). Within the RHRM dimension, the highest number of sample firms reported the highest performance (ahead of, or leading, similar firms) for item 'Improve employee health and safety' at 76.6 %, followed by item 'Creation of good work-life balance and family friendly employment' at 67.1 %. The remaining three RHRM items—'Investor in people', 'Employee participation in decision-making process' and 'Equal opportunities in workplace'—had 53.8 %, 50.6 % and 46.2 % of the sample firms reporting the highest performance levels respectively. In relation to a SSV capability, 47.5 % of the sample firms reported agreement (agreed or strongly agreed) regarding the existence of such a capability in their firm, with only 3.8 % disagreeing (disagree or strongly disagree) about such a capability in their firm. In terms of financial performance, 40.5 % of firms reported better or much better financial performance (compared to similar firms in their industry) for the reporting period.

4.2 Hypotheses Testing

Table 4 shows the means, standard deviations (SD) and correlations among the analysed variables. Estimated correlations between variables were well below the recommended cut-off of 0.7, meaning that discriminant validity was established for the study constructs [83].

To test the research hypotheses, we followed the approach recommended by Baron and Kenny [84]; that is, by constructing separate independent variables for the predictor (EcEnvSoC), moderators (RHRM/SSV), two-way interaction term (multiplying EcEnvSoC with RHRM), and three-way interaction term (multiplying EcEnvSoC with RHRM and SSV). To prove moderation, the interaction term must be significant. There might also be significant main effects for the predictor

Table 5 Results of moderated multiple regression models[1]

	Dependent variable: financial performance			
Variable	Model 1	Model 2	Model 3	Model 4
Constant	3.15***	3.12***	3.12***	3.09***
	(0.08)	(0.07)	(0.07)	(0.04)
Firm size	0.26**	0.17**	0.20**	0.20**
	(0.07)	(0.07)	(0.07)	(0.06)
Perceived GFC	−0.43***	−0.42***	−0.41***	−0.40***
	(0.06)	(0.05)	(0.05)	(0.05)
EcEnvSoC		0.31***	0.21**	0.21**
		(0.06)	(0.07)	(0.07)
RHRM			0.05	0.04
			(0.04)	(0.03)
SSV			0.07	0.05
			(0.06)	(0.04)
EcEnvSoC*RHRM				0.29**
				(0.08)
EcEnvSoC*RHRM*SSV				0.37***
				(0.06)
F	32.79***	28.56***	24.91***	20.39***
Adjusted R^2	0.28	0.44	0.46	0.57
Change in Adjusted R^2		0.16***	0.02	0.11***

Note [1]Values are unstandardised coefficients, with standard errors in parentheses. $^{**}p < 0.01$; $^{***}p < 0.001$

and moderator variables but these effects are not directly relevant conceptually to testing the moderation hypothesis. Unlike the mediator-predictor relation (predictor being an antecedent to the mediator), moderators and predictors are at the same level in regard to their role as variables antecedent or exogenous to certain criterion effects [84].

In this study, we employed the moderated hierarchical regression analysis introducing moderator effects (interaction terms) in the final model [85]. Following the suggestion of previous research [86], we firstly mean-centered both the predictor and moderators prior to creating interaction terms in order to avoid multicollinearity. The variance inflation factor (VIF) for each of the regression coefficients was well below the recommended cut-off of 10 [87], with the VIF values for the final model ranging from 1.02 to 2.04. This result indicated that multicollinearity did not affect the weights of the controls or the hypothesized variables [88].

Table 5 shows the results of moderated multiple regression models. In Model 1, only control variables (firm size and perceived GFC) were included, explaining 28 % of the variance in financial performance ($F = 32.79$, $p < 0.001$). Model 2 introduced the main effect of EcEnvSoC. The result showed that without considering other independent variables, EcEnvSoC had a main positive effect on financial performance, explaining an additional 16 % of the variance in financial performance ($F = 28.56$, $p < 0.001$). RHRM and SSV were entered in Model 3 to examine their main effects.

Fig. 2 Effect of EcEnvSoC on financial performance moderated by RHRM

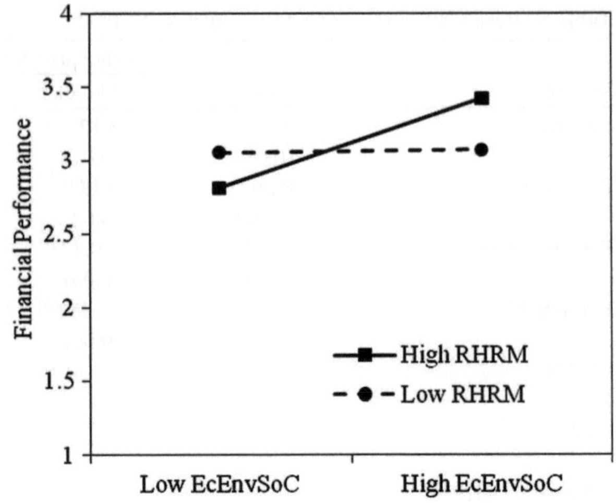

The results indicated that by controlling for the main effect of EcEnvSoC, both RHRM and SSV had no main effect on financial performance, with no significant additional variance explained by adding these two variables (ΔAdjusted $R^2 = 0.02$). In Model 4, the two way interaction term of EcEnvSoC with RHRM and the three-way interaction term of EcEnvSoC, RHRM and SSV were added. The model was significant ($F = 20.39$, $p < 0.001$) and explained 57 % of the variance in financial performance. It also represented a significant improvement over Model 3 as indicated by a change in adjusted R^2 from 0.46 to 0.57 ($p < 0.001$).

As shown for Model 4 of Table 5, the coefficient for EcEnvSoC was significant and positive ($b = 0.21$, $p < 0.01$), indicating the existence of a main positive effect of EcEnvSoC on financial performance. Hence, Hypothesis 1 was supported. The two-way interaction of EcEnvSoC with RHRM was positive and statistically significant ($b = 0.29$, $p < 0.01$), indicating that the benefits of EcEnvSoC for financial performance increased at high levels of RHRM, and thereby providing support for Hypothesis 2. The three-way interaction of EcEnvSoC, RHRM and SSV was positively significant ($b = 0.37$, $p < 0.001$), suggesting that the positive interactive effect of EcEnvSoC and RHRM on financial performance was stronger at high levels of SSV, and thus supporting Hypothesis 3. With regard to the control variables, firm size was found to be positively associated with financial performance ($b = 0.20$, $p < 0.01$), while a negative association was observed between the perceived GFC and financial performance ($b = -0.40$, $p < 0.001$).

In order to explore the nature of these interactions further, we computed simple slopes and graphed the interactions, using high (1 SD above the mean) and low (1 SD below the mean) levels of the moderator. Figure 2 illustrates the effect of EcEnvSoC on financial performance moderated by RHRM, by showing that at high levels of RHRM the association between EcEnvSoC and financial performance was positive and strong (simple slope = 0.41, $p < 0.05$). However, the strength of this association was greatly diminished and became insignificant at low

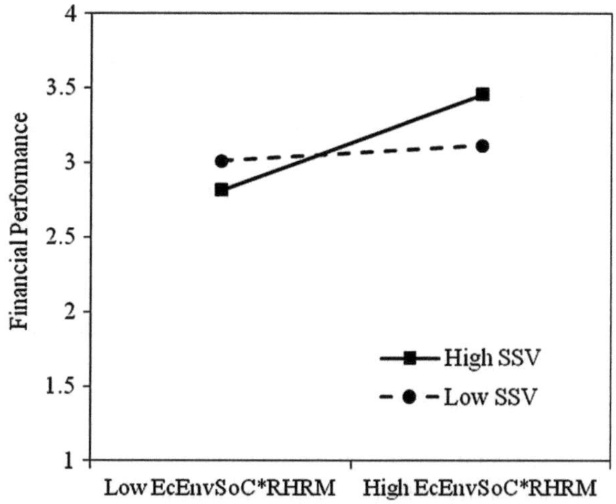

Fig. 3 Effect of the interaction between EcEnvSoC and RHRM on financial performance moderated by SSV

levels of RHRM (simple slope = 0.02, $p = 0.84$), thus providing further support for Hypothesis 2.

Figure 3 shows the effect of the interaction between EcEnvSoC and RHRM on financial performance moderated by SSV. Based on this figure, the association between EcEnvSoC interacting with RHRM and financial performance was positive and strong at high levels of SSV (simple slope = 0.51, $p < 0.01$), but the strength of this association greatly declined and became insignificant at low levels of SSV (simple slope = 0.08, $p = 0.69$), thereby providing further support for Hypothesis 3.

5 Discussion

Our aim for this research has been to examine how and with what impact a firm's HRM practices—considered in terms of responsible HRM practices (RHRM) and a culture-related capability of shared strategic vision (SSV)—can contribute to creating the potential for the other combined elements of CSR (economic and environmental CSR together with social CSR in the community: EcEnvSoC) that result in competitive advantage and superior financial returns for the firm. Consistent with our hypotheses, we have found that RHRM practices and, more importantly the RHRM-SSV interaction, positively moderate the effect of EcEnvSoC on firm financial performance. In other words, the financial benefits of EcEnvSoC increase at high levels of RHRM, and such benefits are likely to be greater when RHRM interacts together with a SSV capability.

The study findings provide empirical evidence for the importance of RHRM practices as fundamental to the achievement of sustainable financial and socially responsible business outcomes for small firms. These findings are in line with previous research [6, 13, 14, 41], and lend support to the argument of Gond and colleagues [52] that RHRM plays a central role as a strategic piece of CSR for sustaining corporate orientation towards a 'triple bottom line'. As suggested by previous research [89], RHRM practices, such as engaging employees and providing values-oriented training and employee development opportunities, can complement environmental-related CSR by building commitment to environmental values and improving the necessary technological skills for adopting such activity. Complementarity may also create an impetus in the firm that drives the transition towards economic-related CSR [90]; properly designed RHRM can trigger innovations that improve production efficiency and lower the cost of a product and that contribute positively to a firm's financial performance.

Our finding in relation to the positive moderating effect of a SSV capability also confirms the importance of the strategic integration of collective values about organisational goals and of the exploitation of new business opportunities in generating a better effective implementation of RHRM that promotes financial performance in small firms. This finding is consistent with other previous empirical and theoretical research relevant to the SME context [25, 26, 64, 78], as well as in line with a primary tenet in RBV theory that a firm's capabilities are fundamental drivers of competitive advantage and superior performance [54].

It is also of interest to note—based on our descriptive statistics showing the 'moderate-to-high' RHRM adoption rate among the sample firms (nearly 60 % reporting ahead of/leading similar firms on RHRM overall)—that small firms are more likely to embrace the adoption of RHRM (albeit perhaps on an ad hoc basis, rather than as part of a deliberate strategic approach) more widely than might be expected given the size-related resource constraints under which many operate [91, 92, 94–96]. In this study, the weight given to RHRM by the majority of firms is found to be greater than that given to performance in regard to economic-related CSR practices. This result may reflect recognition of the importance of employees to the achievement of a firm's business objectives generally [12], and lends support to the notion that the value of CSR becomes debased and only an exercise in public relations when employees are not fully engaged [97].

By focusing on specific types of RHRM, 'employee health and safety' and 'work-life balance and family friendly employment' have the highest performance levels reported by the majority of the sample firms (77 % and 67 % respectively). These results may reflect the Australian institutional profile (see [98] for extensive discussion of the 'institutional profile' concept) where the existing industrial regulatory framework, the values/beliefs about good management practices held by owner-managers, and a prevalence of ad hoc arrangements developed in response to requests from individual employees, make it more common for small-firm workplaces in Australia to actively implement and promote these certain types of RHRM.

The descriptive analysis of our survey data also indicates that only slightly more than half of firms report the highest performance levels in 'investing in employees' and 'employee participation in decision-making' (54 % and 51 % respectively), whilst less than half of them perform at the highest levels in terms of 'equal opportunities' (46 %). Taking investment in employees first, although there is no agreement in the literature regarding the level of training provided by small firms [99], our finding of a moderate adoption rate accords with published research in the UK context [5, 92], which may reflect the difficulties of measuring 'informal' training provision (e.g. on-the-job training) in small firms causing the true level of training provision to be underreported [100].

In relation to employee participation, although our finding of a 'moderate-to-low' adoption rate aligns with the results of previous research [10] that Australian SMEs have been reluctant to use management practices that support employee participation, it would be misleading to interpret this finding without considering the existence of a SSV capability—which relates to employees being able to influence the way to work and the objectives of the firm, and to freely contribute their points of view about how to manage the firm smoothly—in the sample firms. Evidence presented in this study of only 4 % of firms reporting disagreement regarding the existence of such a capability in their firms, suggest that the principle of employee participation is, in fact, highly accepted amongst small firms in our sample, but they might not have 'formal' employee-participation policies in place. This explanation is supported by the findings from the 2004 Workplace Employment Relations Survey (WERS) in the UK indicating that in contrast to employees in medium and large firms, the majority of employees in small firms report high satisfaction (good or very good) with their workplace manager in keeping them informed about, and participating in, changes to the running of the business, changes in staffing and changes in their job [99].

Lastly, whilst on the one hand the low rate of adopting equal opportunity practices found in this study may indicate that such practices are less common in workplaces in Australian small firms, on the other hand it may reflect a lack of formal written equal opportunity policies among the sample firms to carry out a range of monitoring activities to assess whether employment practices have any discriminatory effects. Our assumption is confirmed by the results from the 2007 WERS showing that workplaces in smaller firms make less use of equal opportunity practices than those in larger firms due to a lack of policy formality, but even so, employees in smaller firms are more likely to experience fair treatment in their workplace than those in larger firms [99].

6 Conclusions

The study has significant practical management implications. The empirical evidence presented in this study highlights the critical role of responsible HRM practices plus a culture-related capability in successful implementation of CSR that promotes financial performance of small firms. It appears, based on our findings, that the borders of HRM and CSR are increasingly blurred [52], where responsible HRM practices reported by small firms are informed by the principles of CSR and the need to enhance financial performance while creating the sustainable outcomes of value increasingly required of business. In particular, the study findings indicate the need for managers, who wish their small firm to remain financially competitive from their CSR efforts, to be aware of the importance of implementing CSR policies with RHRM practices that reflect and fulfill employees' needs and which avoid adopting a paternalistic approach that neglects the employees' perspective. It is also worthwhile to note that unlike large firms, there seems to be a lack of formality in HRM practices in small firms that provide an important benefit. Workplaces which are part of small firms tend to have high levels of informality and familial ties that can help to engender trust and reciprocity in manager-employee interactions.

In common with most research, this study has some limitations. First, the sample size was small which could reduce the reliability and stability of parameter estimates. Research suggests that the detection of moderating effects is often constrained by small sample size [101]. However, given that all of our hypotheses were supported, we can be confident in the robustness of our findings. Second, difficulties in generalizing results to other industry sectors and to other economies should be acknowledged. Third, the findings are limited by the self-reported nature of the data collection process. Finally, although we used the 6-month time lag in data collection to allow for temporal ordering of independent/moderating and dependent variables, this time gap was insufficient time to confirm deterministic causation. Future research might include a quasi-experimental longitudinal study over a longer time period with a larger sample size to evaluate the RHRM-SSV interaction effects with CSR on financial performance, thereby allowing for broader generalizability in findings. Linking the model variables to multiple data sources that take into account objective measures would further extend our understanding of how responsible HRM practices and capabilities could strengthen the CSR-performance relationship.

References

1. Dunphy, D. (2003). Corporate sustainability: Challenge to managerial orthodoxies. *Journal of the Australian and New Zealand Academy of Management, 9*(1), 2–11.
2. Jenkins, H. (2009). A business opportunity model of corporate social responsibility for small-and medium-sized enterprises. *Business Ethics: A European Review, 18*(1), 21–36.
3. Porter, M. E., & Kramer, M. R. (2006). Strategy and society: The link between competitive advantage and corporate social responsibility. *Harvard Business Review, 84*(12), 78–92.
4. Gollan, P. J. (2000). Human resources, capabilities, and sustainability. In D. Dunphy, J. Benveniste, A. Griffiths, & P. Sutton (Eds.), *Sustainability: The corporate challenge of the 21st century* (pp. 55–77). Sydney: Allen and Unwin.
5. Hughes, M., Keddie, V., Webb, P., & Corney, M. (2002). *Working towards skills: perspectives in workforce development in SMEs*. London: Learning and Skills Development Agency.
6. Roehling, M. V., Boswell, W. R., Caligiuri, P., Feldman, D., Graham, M. E., Guthrie, J. P., et al. (2005). The future of HR management: Research needs and directions. *Human Resource Management, 44*(2), 207–216.
7. Stone, R. (2011). *Human resource management* (7th ed.). Milton: Willey.

8. Barrett, R., & Buttigieg, D. (1999). Trade unions and small firms: Is it size that counts? *Labour and Industry, 9*(3), 41–55.
9. Bartram, T. (2005). small firms, big ideas: The adoption of human resource management in australian small firms. *Asia Pacific Journal of Human Resources, 43*(1), 137–154.
10. Wiesner, R., & McDonald, J. (2001). Bleak house or bright prospect? Human resource management in Australian SMES. *Asia Pacific Journal of Human Resources, 39*(2), 31–53.
11. Heneman, R. L., Tansky, J. W., & Camp, S. M. (2000). Human resource management practices in small and medium-sized enterprises: Unanswered questions and future research perspectives. *Entrepreneurship Theory and Practice, 25*(1), 11–26.
12. Mankelow, G. (2008). Social responsibility paradox of small business human resource management practices. *The International Journal of Human Resource Management, 19*(12), 2171–2181.
13. Gollan, P. J. (2006). Sustainable strategies towards human resources: A way forward. In D. Marinova, D. Annandale, & J. Phillimore (Eds.), *International handbook on environmental technology management* (pp. 104–118). London: Edward Elgar.
14. Jabbour, C. J. C., & Santos, F. C. A. (2008). Relationships between human resources dimensions and environmental management in companies: Proposal of a model. *Journal of Cleaner Production, 16*, 51–58.
15. Martin-Tapia, I., Aragon-Correa, J. A., & Llamas-Sanchez, R. (2008). The relationship between high performance work systems and proactive environmental management. In R. Wustenhagen, J. Hamschmidt, S. Sharma, & M. Starik (Eds.), *Sustainable innovation and entrepreneurship* (pp. 197–224). Cheltenham: Edward Elgar.
16. Vickers, M. (2005). Business ethics and the HR role: Past, present, and future. *Human Resource Planning, 28*(1), 26–32.
17. Montiel, I. (2008). Corporate social responsibility and corporate sustainability: Separate pasts, common futures. *Organization & Environment, 21*(3), 245–269.
18. Commission, European. (2003). *Responsible entrepreneurship: A collection of good practice cases among small and medium-sized enterprises across Europe*. Luxembourg: European Commission.
19. Aragon-Correa, J. A., & Sharma, S. (2003). A contingent resource-based view of proactive corporate environmental strategy. *Academy of Management Review, 28*(1), 71–88.
20. Benn, S., Dunphy, D., & Griffiths, A. (2006). Enabling change for corporate sustainability: An integrated perspective. *Australasian Journal of Environmental Management, 13*(3), 156–165.
21. Du, S., Bhattacharya, C. B., & Sen, S. (2007). Reaping relational rewards from corporate social responsibility: The role of competitive positioning. *International Journal of Research in Marketing, 24*(3), 224–241.
22. Groza, M. D., Pronschinske, M. R., & Walker, M. (2011). Perceived organizational motives and consumer responses to proactive and reactive CSR. *Journal of Business Ethics, 102*, 639–652.
23. Klassen, R. D., & Whybark, D. C. (1999). The impact of environmental technologies on manufacturing performance. *Academy of Management Journal, 42*(6), 599–615.
24. Sharma, S., & Vredenburg, H. (1998). Proactive corporate environmental strategy and the development of competitively valuable organizational capabilities. *Strategic Management Journal, 19*(8), 729–753.
25. Torugsa, N. A., O'Donohue, W., & Hecker, R. (2012). Capabilities, proactive CSR and financial performance in SMEs: Empirical evidence from an Australian manufacturing industry sector. *Journal of Business Ethics, 109*(4), 483–500.
26. Torugsa, N. A., O'Donohue, W., & Hecker, R. (2013). Proactive CSR: An empirical analysis of the role of its economic, social and environmental dimensions on the association between capabilities and performance. *Journal of Business Ethics, 115*(2), 383–402.
27. Bansal, P. (2005). Evolving sustainably: A longitudinal study of corporate sustainable development. *Strategic Management Journal, 26*(3), 197–218.

28. Dyllick, T., & Hockerts, K. (2002). Beyond the business case for corporate sustainability. *Business Strategy and the Environment, 11*(2), 130–141.
29. Willard, B. (2005). *The sustainability advantage: Seven business case benefits of a triple bottom line*. Canada: New Society Publisher.
30. Walley, N., & Whitehead, B. (1994). It's not easy being green. *Harvard Business Review, 72*(3), 46–52.
31. Aragon-Correa, J. A. (1998). Strategic proactivity and firm approach to the natural environment. *Academy of Management Journal, 41*(5), 556–567.
32. Brammer, S., & Millington, A. (2006). Firm size, organizational visibility and corporate philanthropy: An empirical analysis. *Business Ethics: A European Review, 15*(1), 6–18.
33. Lepoutre, J., & Heene, A. (2006). Investigating the impact of firm size on small business social responsibility: A critical review. *Journal of Business Ethics, 67*(3), 257–273.
34. Jenkins, H. (2006). Small business champions for corporate social responsibility. *Journal of Business Ethics, 67*(3), 241–256.
35. Harrison, J. G., & Freeman, R. E. (1999). Stakeholders, social responsibility and performance: Empirical evidence and theoretical perspectives. *Academy of Management Journal, 42*(5), 479–485.
36. Hart, S. L., & Ahuja, G. (1996). Does it pay to be green?: An empirical examination of the relationship between emission reduction and firm performance. *Business Strategy and the Environment, 5*, 30–37.
37. Mackey, A., Mackey, T. B., & Barney, J. B. (2007). Corporate social responsibility and firm performance: Investor preferences and corporate strategies. *Academy of Management Review, 32*(3), 817–835.
38. Hammann, E., Habisch, A., & Pechlaner, H. (2009). Values that create value: Socially responsible business practices in SMEs—empirical evidence from German companies. *Business Ethics: A European Review, 18*(1), 37–51.
39. Jamrog, J., Vickers, M., & Bear, D. (2006). Building and sustaining a culture that supports innovation. *Human Resource Planning, 29*(3), 9–19.
40. Matthews, J. (2002). Innovation in Australian small to medium enterprises: Contributions from strategic human resource management. *Asia Pacific Journal of Human Resources, 40*(2), 193–204.
41. Chew, I. K. H., & Sharma, B. (2005). The effects of culture and HRM practices on firm performance: Empirical evidence from Singapore. *International Journal of Manpower, 26*(6), 560–581.
42. Laursen, K., & Foss, N. J. (2003). New HRM practices, complementarities, and the impact on innovation performance. *Cambridge Journal of Economics, 27*(2), 243–263.
43. Fernandez, E., Junquera, B., & Ordiz, M. (2003). Organizational culture and human resources in the environmental issue: A review of the literature. *The International of Human Resource Management, 14*(4), 634–656.
44. Rothenberg, S. (2003). Knowledge content and worker participation in environmental management at NUMMI. *Journal of Management Studies, 40*(7), 1783–1802.
45. Berry, M. A., & Rondinelli, D. A. (1998). Proactive corporate environment management: A new industrial revolution. *Academy of Management Executive, 12*(2), 38–50.
46. Bianchi, R., & Noci, G. (1998). Greening SMEs' competitiveness. *Small Business Economics, 11*(3), 269–281.
47. Ramus, C. A., & Steger, U. (2000). The roles of supervisory support behaviors and environmental policy in employee 'Ecoinitiatives' at leading-edge European companies. *Academy of Management Journal, 43*(4), 605–626.
48. Greenwood, M. R. (2002). Ethics and HRM: A review and conceptual analysis. *Journal of Business Ethics, 36*, 261–278.
49. Legge, K. (1998). The morality of HRM. In C. Mabey, D. Skinner, & T. Clark (Eds.), *Experiencing human resource management* (pp. 14–32). London: Sage.
50. Winstanley, D., Woodall, J., & Heery, E. (1996). Business ethics and human resource management themes and issues. *Personnel Review, 25*(6), 5–12.

51. Shen, J., & Zhu, C. J. (2011). Effects of socially responsible human resource management on employee organizational commitment. *The International Journal of Human Resource Management, 22*(15), 3020–3035.
52. Gond, J. P., Igalens, J., Swaen, V., & Akremi, A. E. (2011). The human resource contribution to responsible leadership: An exploration of the CSR-HR interface. *Journal of Business Ethics, 98*, 115–132.
53. Barney, J. B. (1986). Types of competition and the theory of strategy: Toward an integrative framework. *Academy of Management Review, 11*(4), 791–800.
54. Barney, J. B. (1991). Firm resources and sustained competitive advantage. *Journal of Management, 17*(1), 99–120.
55. Andersson, L. M., & Bateman, T. S. (2000). Individual environmental initiative: Championing natural environmental issues in US business organizations. *Academy of Management Journal, 43*, 548–570.
56. Hart, S. L. (1995). A natural-resource-based view of the firm. *Academy of Management Review, 20*(4), 986–1014.
57. Buysse, K., & Verbeke, A. (2003). Proactive environmental strategies: A stakeholder management perspective. *Strategic Management Journal, 24*(5), 453–470.
58. Christmann, P. (2000). Effects of 'best practices' of environmental management on cost advantage: the role of complementary assets. *Academy of Management Journal, 43*(4), 663–680.
59. Sharma, S., Aragon-Correa, J. A., & Rueda-Manzanares, A. (2007). The contingent influence of organizational capabilities on proactive environmental strategy in the service sector: An analysis of north american and european ski resort. *Canadian Journal of Administrative Science, 24*(4), 268–283.
60. Marcus, A. A., & Geffen, D. (1998). The dialectics of competency acquisition: Pollution prevention in electric generation. *Strategic Management Journal, 19*, 1145–1168.
61. Boxall, P., & Purcell, J. (2003). *Strategy and human resource management*. Hampshire: Palgrave Macmillan.
62. Tsai, W., & Ghoshal, S. (1998). Social capital and value creation: The role of intrafirm networks. *Academy of Management Journal, 41*(4), 464–476.
63. Campbell, A., & Yeung, S. (1991). Creating a sense of mission. *Long Range Planning, 24*(4), 10–20.
64. Pfeffer, J. (2005). Changing mental models: HR's most important task. *Human Resource Management, 44*(2), 123–128.
65. Roberts, R., & Hirsch, P. (2005). Evolution and revolution in the twenty-first century: Rules for organizations and managing human resources. *Human Resource Management, 44*(2), 171–176.
66. Chen, M., & Hambrick, D. C. (1995). Speed, stealth, and selective attach: How small firms differ from large firms in competitive behavior. *Academy of Management Journal, 38*(2), 453–482.
67. Covin, J. G., & Slevin, D. P. (1990). New venture strategic posture, structure and performance: An industry life cycle analysis. *Journal of Business Venturing, 5*(2), 123–135.
68. Veliyath, R., & Shortell, S. M. (1993). Strategic orientation, strategic planning system characteristics and performance. *Journal of Management Studies, 30*(3), 359–381.
69. Boudreau, J. W., & Ramstad, P. M. (2005). Talentship, talent segmentation, and sustainability: A new HR decision science paradigm for a new strategy definition. *Human Resource Management, 44*(2), 129–136.
70. Dun and Bradstreet. (2009). *Australia's premier source of business intelligence just got even better*. Australia: Dun and Bradstreet.
71. Chandler, G. N., & Hanks, S. H. (1993). Measuring the performance of emerging business: A validation study. *Journal of Business Venturing, 8*(5), 391–408.
72. Lyon, D. W., Lumpkin, G. T., & Dess, G. G. (2000). Enhancing entrepreneurial orientation research: Operationalizing and measuring a key strategic decision making process. *Journal of Management, 26*(5), 1055–1085.

73. Gardner, T. M., & Wright, P. M. (2009). Implicit human resource management theory: A potential threat to the internal validity of human resource practice measures. *The International Journal of Human Resource Management, 20*(1), 57–74.
74. Podsakoff, P. M., MacKenzie, S. M., Lee, J., & Podsakoff, N. P. (2003). Common method variance in behavioral research: A critical review of the literature and recommended remedies. *Journal of Applied Psychology, 88*(5), 879–903.
75. Podsakoff, P. M., & Organ, D. W. (1986). Self reports in organizational research: Problems and prospects. *Journal of Management, 12*(4), 531–544.
76. Pierce, J. L., Gardner, D. G., Dunham, R. B., & Cummings, L. L. (1993). Moderation by organization-based self-esteem of role condition-employee response relationships. *Academy of Management Journal, 36*(2), 271–288.
77. Armstrong, J., & Overton, T. (1977). Estimating nonresponse bias in mail surveys. *Journal of Marketing Research, 14*, 396–402.
78. Aragon-Correa, J. A., Hurtado-Torres, N., Sharma, S., & Garcia-Morales, J. V. (2008). Environmental strategy and performance in small firms: A resource-based perspective. *Journal of Environmental Management, 86*(1), 88–103.
79. Dess, G. G., & Robinson, R. B. (1984). Measuring organizational performance in the absence of objective measures: The case of the privately-held firm and conglomerate business unit. *Strategic Management Journal, 5*(3), 265–275.
80. Homburg, C., Krohmer, H., & Workman, J. P. (1999). Strategic consensus and performance: The role of strategy type and market-related dynamism. *Strategic Management Journal, 20*(4), 339–357.
81. Ansoff, H. I. (1965). *Corporate strategy*. New York: McGraw-Hill.
82. Lin, H. P., Huang, W. C., Chen, H. F., & Ke, Y. P. (2011). An empirical study of Taiwan's hospital foundation investment in corporate social responsibility and financial performance. *World Academy of Science, Engineering and Technology, 78*, 345–349.
83. Pallant, J. (2007). *SPSS: survival manual: A step-by-step guide to data analysis using SPSS for windows, Version 15*. New South Wales: Allen & Unwin.
84. Baron, R. M., & Kenny, D. A. (1986). The moderator-mediator variable distinction in social psychological research: Conceptual, strategic and statistical considerations. *Journal of Personality and Social Psychology, 51*, 1173–1182.
85. Cohen, J., & Cohen, P. (1983). *Applied multiple regression/correlation analysis for the behavioural sciences*. Hillsdale: Lawrence Erlbaum.
86. Aiken, L. S., & West, S. G. (1991). *Multiple regression: Testing and interpreting interactions*. Newbury Park: Sage.
87. Mason, C., & Perreault, W. (1991). Collinearity, power, and interpretation of multiple regression analysis. *Journal of Marketing Research, 28*, 268–280.
88. Neter, J., Wasserman, W., & Kutner, M. H. (1985). *Applied linear statistical models: Regression, analysis of variance, and experimental design*. Homewood: Richard D. Irwin.
89. Graafland, J., van de Ven, B., & Stoffele, N. (2003). Strategies and instruments for organising CSR by small and large businesses in the Netherlands. *Journal of Business Ethics, 47*(1), 45–60.
90. Chang, D., & Kuo, L. R. (2008). The effects of sustainable development on firms' financial performance—an empirical approach. *Sustainable Development, 16*, 365–380.
91. Bacon, N., Ackers, P., Storey, J., & Coates, D. (1996). 'It's a small world: Managing human resources in small business. *The International Journal of Human Resource Management, 7*, 82–100.
92. Cassell, C., Nadin, S., Gray, M., & Clegg, C. (2002). Exploring human resource management practices in small and medium sized enterprises. *Personnel Review, 31*, 671–692.
93. Cohen, J. (1988). *Statistical power analysis for the behavioral sciences*. Hillsdale: Lawrence Erlbaum.
94. de Kok, J., & Uhlaner, L. (2001). Organisation context and human resource management in the small firm. *Small Business Economics, 17*, 273–291.

95. Duberley, J., & Walley, P. (1995). Assessing the adoption of HRM by small and medium sized manufacturing organisations. *The International Journal of Human Resource Management, 6*, 891–909.
96. Wiesner, R., & Innes, P. (2010). Bleak house or bright prospect? Human resource management in Australian SMEs over 1998–2008. *Asia Pacific Journal of Human Resources, 48*, 151–183.
97. Mees, A., & Bonham, J. (2004). Corporate social responsibility belongs with HR. *Canadian HR Reporter, 17*(7), 11–13.
98. Kostova, T., & Roth, K. (2002). Adoption of an organizational practice by subsidiaries of multinational corporations: Institutional and relational effects. *Academy of Management, 45*(1), 215–233.
99. Forth, J., Bewley, H., & Bryson, A. (2004). *Small and medium-sized enterprises: Findings from the 2004 workplace employment relations survey*. London: Department of Trade and Industry.
100. Patton, D., Marlow, S., & Hannon, P. (2000). The relationship between training and small firm performance: Research frameworks and lost quests. *International Small Business Journal, 19*(1), 11–27.

Managing High Performance Work Systems and Organizational Performance

Aykut Berber and Murat Yaslioglu

Abstract Workplace is primarily a social environment and value is created not by machines or computers but by people who operate them. Thus, high performance work systems should allow individuals to experience greater autonomy over their tasks and particularly self-managed teams are useful where a group of individuals can all together plan, organize and control their tasks without the need for micro-management. However, forming up and maintaining self-managed teams to achieve highest performance has always been a managerial concern but has never been easy.

1 The Essence of Organizing Work for High Performance

In an article published towards the end of the last century, Kling was drawing our attention to the competitiveness of the world economy: While adapting themselves to the changing conditions, American companies were more concerned with the product quality. As an outgrowth of these new dynamics, 'creativity', 'ingenuity' and 'problem solving abilities' became priority issues for the survival of these companies. Having carefully examined the conditions of the day, Kling put particular emphasis on the employees; they needed to be provided with 'information', 'incentives' and 'responsibilities to make decisions for innovation, quality improvement, and rapid response to change'. Companies taking these issues into consideration were referred to as—according to the author—'high performance work organizations' [1].

A. Berber (✉) · M. Yaslioglu
Department of Management and Organization, Istanbul University
School of Business, 34820 Istanbul, Avcilar, Turkey
e-mail: berber@istanbul.edu.tr

C. Machado and J. P. Davim (eds.), *Work Organization and Human Resource Management*, Management and Industrial Engineering,
DOI: 10.1007/978-3-319-06376-8_2,

1.1 'Citius, Altius, Fortius' *in Business*

Kling's words are, of course, one of many similar opinions declared by management scholars in 1990s, and a retrospective view towards those years would prove that most of them—if not all—were right. Companies in developed and developing economies during the last decade of the twentieth century were experimenting this transition—a challenging period where (i) customers were getting more involved in product development efforts (ii) being part of a supply network was preferred over depending on a strictly controlled supply chain, and (iii) innovation became the motto for product differentiation and even for taking place in uncontested markets. Specifically for the first argument, a research conducted in 1990 by Zirger and Maidique revealed that 'managerial excellence' was highly critical to the product success, while a new product was meant to provide 'significant value to the customer' [2]. Products that were unique, or at least that were remarkably different in the market, were substantial to the company's long-term success, but customers, on the other hand, were precious. A strong product concept, as stated by Clark and Fujimoto in 1990, was crucial; their development process was 'a complex system for solving problems and making decisions'. However, according to the authors, organizing for integrity was as essential as the product integrity. Spending efforts on products with distinctive features that responded to the needs and requirements of the customers, who anticipated new products in line with their values and lifestyles, and integrating these efforts into the development process were also the major concerns of companies of the particular era [3].

When compared to today's circumstances, mid-1990's were far less ambiguous but served as a threshold before the gate into a new age for the business world. The Internet was only just introduced at the service of the public but its use in business organizations was rapidly increasing. As a matter of fact, even this little efflorescence of the knowledge age caused great excitement among management scholars, who began proposing countless ways for responding quickly as well as accurately to the *exact* needs and requirements of the customers. For example, Angell and Heslop's book on 'growing the business in the electronic age' excellently described everything necessary to assess a company's needs and opportunities for a better exploitation of the Internet and developing online strategies [4]. Following a newly built avenue in the field of post-modern management theories, knowledge management and learning organizations were regarded as providers of new and efficient formulas in business organizations to enliven people's souls at work and motivate them (e.g. [5–7]). A revolutionary approach began to sail before the wind in the seas of the human resource management. In 1998, Ulrich set forth the following noteworthy formula [8]:

$$\textit{Intellectual capital} = \textit{Competence} \times \textit{Commitment}$$

Ulrich's article was quite refreshing; it revealed five particular tools in order to increase competence in a business organization: (i) 'hiring new talents' (ii) 'investing in employee learning and training' (iii) 'organizing partnerships with suppliers, customers and vendors' so that knowledge could be shared, new knowledge could be created, new work methods could be brought into the organization' (iv) 'removing employees who fail to change, learn and adapt', and (v) 'keeping successful employees'. Concurrently with these efforts; empowering employees in terms of decision-making and participation, encouraging teamwork, sharing long-term information and strategies with employees, providing new technologies were only a few of the several initiatives that should be taken by the managers in order to enhance commitment [8]. On the other hand, in 1997, Baldwin and colleagues put an emphasis on the issue of 'learning', which could 'lead an organizations' response to changes within its business environment'. Their model remarkably brought a refreshing interpretation of the importance of learning: As the environmental turbulence increased, suggested the authors, the strategic role of learning evolved from a focus on 'employee development' into two consecutive stages—a focus on 'imminent business needs', and a focus on 'unknown business development'. Predictable environmental conditions were considered easy to manage; managers could focus on the development of their employees' individual knowledge in already existing business practices. However, as the environment became more complicated, necessary actions would be essentially taken from focusing on customer requirements and innovation in existing business practices to focusing on undefined market potential and assessments of changes to decide what the business should become [9]. The final stage apparently required managers to think beyond the boundaries of their organizations and to exploit diverse opportunities and resources.

Defending the fact that the managerial point of view had often been the point of origin when describing the evolution of management, in 1996, Gollac and Volkoff conceptualized the accelerated work life and used the expression *l'intensification du travail* to describe the context of working harder and faster. Their arguments were quite elucidative—'technical and organizational progress' were not sufficient enough to bring hard work to an end. Companies began to demand more (… and more) from their employees, and they were looking for 'better-trained employees' while practicing 'restrictive salary policies'. Just as the title of Gollac and Volkoff's study goes, the managerial situation in organizations had simply turned into a '*citius, altius, fortius*' fashion [10]. Being originally a Latin expression for 'swifter, higher, stronger', '*citius, altius, fortius*' is the motto of the Olympic Games coined by a liberal and controversial priest, Henri Didon, and introduced by the founder of the International Olympic Committee, Pierre de Coubertin to delineate the Olympic movement more than a century ago (p. 128 in [11]). Sports is generally a physical activity; competition among individuals or teams is encouraged to show highest performance possible, break records and play at full capacity to win games according to the rules that are previously set.

1.2 Searching For Excellence: A Dream or an Obsession?

First published in 1982, Peters and Waterman have introduced to the business world a ground-breaking book, *In Search of Excellence*, which is assumed to be one of the most widely read business books ever. Peters and Waterman delicately and confidently put the *factual* conditions in the business environment into words based on three following postulates (Authors' Note in [12]):

1. People and organizations are *not* rational; therefore, managing just 'by numbers' is erroneous.
2. People are far beyond mere factors of production; treating people as cogs in a machine is demotivating and must be avoided.
3. The world is full of ambiguity.

Calling attention to these conditions, the two notable authors revealed eight 'attributes of excellence' that distinguished successful companies, and these attributes varied from taking action and learning from failures to keeping in close touch with customers; managing toward a value system; arranging simple and workable organizational structures; and delivering people the freedom of doing things their own way, while practicing a centralized management of such essential elements that needed careful supervision as 'one or two top strategic priorities', 'central values that make up the company's culture', and some 'key financial indicators'. A little more than twenty years later, in the reprint edition of their book, Peters and Waterman reframed their opinion about these attributes in an additional section before the preface called *Authors' Note: Excellence 2003*. In essence, they did not seem to have changed their mind—these eight attributes were still 'their eight attributes of excellence'. Having written numerous titles since 1980s, both authors enunciated their determination on sticking with these attributes (Authors' Note in [12]). In fact, *In Search of Excellence* was a seminal work that had repercussions. It was acclaimed by many scholars with great enthusiasm and served as basis to many studies. However, a great deal of critical writing was also evident (e.g. [13–15]). A debate on whether these eight attributes of excellence were genuinely (or scientifically) effectual may be considered a separate matter of discussion, which is out of our immediate scope. Nevertheless, using such a word—which denotes a level of quality that surpasses ordinary standards—as *excellence* seems quite tempting if one wants to open a ceaseless discussion crowded with alternative perspectives and opposing ideas in the field of management.

1980s and 1990s witnessed a revolutionary transition process underlying management thinking. *Organizational change* was perhaps the hot-button issue of the era and *change management* became the topic of endless discussions. If companies wanted to survive, or maintain their long-term existence in the markets, managers had to compromise. Successful managers were not perceived as *savoir-faire* type of heroes anymore. Power and knowledge had to be shared across the organization; and authority and responsibility had to be delegated with employees. Unilateral approaches of managers were considered archaic in successful companies. Quantified targets, or performance ratings were certainly still in vigor but operational goals, objectives and targets were determined in chorus so as to take employees' performances also into consideration. Particularly middle managers began to assume new responsibilities, and as explicitly highlighted by Bradford and Cohen in the introduction to the paperback edition of their book that was published in 1997 and called *Managing for Excellence*, managers gained on the clear conception of their dependency on the knowledge and abilities of their subordinates in order to achieve excellence. In the era of the 'post-heroic' management, managers were required to act as leaders, and henceforth, these leaders were responsible to create circumstances for subordinates where they could show superb performances (pp. v-ix in [16]). Bradford and Cohen's assessment of a fundamental problem encountered in the organizations of the era was quite a subtlety: The middle and upper-middle managers were the ones who were 'holding the key to high performance' since their current position was a striking example to the 'underutilization of human resources'. Having experienced ordinary and routine tasks in the beginning of their careers, middle managers, in particular, were overwhelmed with responsibilities of transforming the sublime strategic plans issued by top managers into departmental plans that were understandable, feasible and doable by their subordinates. Yet, subordinates were not committed enough to the manager's overall department. Their narrow focus was limited to the scope of their job descriptions and on these grounds; the critical real-time information or potential problems were not shared effectively even though the subordinates were usually aware of them. As put forth by Bradford and Cohen, Ouchi's *Theory Z and* Pascale and Athos' *The Art of Japanese Management in 1981*, Deal and Kennedy's *The Corporate Cultures and* Peters and Waterman's *In Search of Excellence in 1982*, or Kanter's *The Change Masters in 1983; they* were all focusing on the total organization (pp. 2–6 in [16]) and [12, 17–20]. The managers in the new era intended to realign the cumbersome organizational structures, to revolutionize the workplace, and to put a new leadership style into practice that would guide employees for the use of their full—and real—capacities to achieve higher performance at work so as to introduce value added products to the markets. At this juncture, many questions follow: Excellence needed strict targets, but who was to put these targets? And once achieved, as whose success was it going to be declared and celebrated? The manager, or the employee?

1.3 High Performance Work Systems

Effective human resource practices definitely lead companies to considerable competitive advantages (e.g. [21, 22]); and taking this axiom into consideration as the point of departure, Ramsay and his colleagues called attention to the notion of a 'high road approach to management'—an approach (or a paradigm) that highly counted on human resource development and praised employee contributions in order to maintain an organization's competitiveness in the market. Reconciling these two points of view—human resource development and competitiveness in the market—the apparent role of high performance work systems, where advanced human resource management practices are combined in multiple ways [21, 23], might be a good point of reference to explicate the issue through a couple of examples. In 1974, departing from Braverman's argument of 'the Taylorist deskilling and direct forms of managerial control as the key mechanisms of work intensification', and being inspired by *neo-Fordist* theory, Ramsay and colleagues stated that 'Taylorism had limits', and the work organization (of the era) tend to quit the Taylorist approach of direct control [21, 24, 25]. At this point, an intention to reorganize the work system should not be confused with a radical intention of breaking away from such fundamental pillars of the *conventional work system* as the division of labor and specialization, or abolishing the hierarchy and managerial levels; it should rather be considered as attempts to overcome certain rigidities within the work processes. Intensification of work is still as an outstanding matter; and the use of new technologies and task rearrangement in line with contemporary patterns in the business environment are appreciated to achieve high performance at work [26–28].

Performance can simply be defined as 'the extent to which an individual contributes to achieving the goals and objectives of an organization' (p. 155 in [29]). However, this simple definition embodies an excessive amount of factors and apparently minimizes the opportunity of a concrete definition for high *performance*. According to Kirby, the source of challenge relies on such a variety of dimensions as holding the highest market share, having the greatest market growth or long-term existence in the market; however, a consensus theory of high performance is hopefully possible as long as studies continue to reveal the secrets behind this multi-faceted problem [30]. A content rich explanation for high performance work was illustrated by Evans and Lindsay in 2012. According to these scholars, 'high performance work was characterized by flexibility, innovation, knowledge and skill sharing, alignment with organizational directions, customer focus, and rapid response to changing business needs and marketplace requirements' (pp. 155–156 in [29]). At this point, teamwork practices usually serve as the principal part of the whole picture—e.g. *cross-functional teams* where diverse ideas from a variety of sources are deliberately brought together to create a new product or a process; *quality circles* where a group of individuals willingly spend some time on problem-seeking and solving efforts in a proactive manner; or *self-managed teams* where a group of individuals attempt to combine their respective skills to plan, organize and control their tasks almost under no direct supervision. An individual employee participating in either of these practices, by all means, are anticipated to be aware of such conditions and values of the contemporary workplace as the significance of sharing knowledge with others, the severity of responding to the most—if not all—of the customer needs, or the necessity of exploiting training opportunities both for self-development and for the organization's interest.

Nearly a decade before Kirby's affirmative article, Gephardt and Van Buren stated that there was no consensus on a definition of a high performance work system—scholars had not agreed upon its components neither [31]. Although the general outlook of the prospective theory seemed to have evolved into a *jungle* (in memoriam Harold Koontz referring to his description of management theory in the early 1960s), works of scholars over the past years shared common ideas; and thanks to the mindful work of Kirkman and his colleagues in 1999, five components 'believed to be representative of all the definitions' made by many scholars illustrate an organized big-picture. After presenting several definitions particularly proposed by leading scholars studying high performance work organizations; Kirkman and colleagues revealed their integrated definition comprised of five components: (i) 'self-managing work teams'; (ii) 'employee involvement, participation, empowerment'; (iii) 'total quality management'; (iv) integrated production technologies'; and (v) 'the learning organization' (pp. 2–3 in [32]).

If, as stated by Appelbaum in 2000, 'the core of a high-performance work system … is that work is organized to permit front-line workers to participate in decisions that alter organizational routines' (p. 7 in [33]), problem-solving and quality improvement efforts may only be possible by making good use of opportunities that appear as an outcome of interpersonal relations. Workplace is primarily a social environment and value is created not by machines and computers but by people who operate them. High performance work systems should allow individuals to experience greater autonomy over their tasks whereupon an efficient and effective communication about issues and problems at work with other individuals will become evident. Participating in decisions, digital literacy and contemporary incentive systems require qualified, experienced, well-trained and, without question, highly talented people with social and firm-specific skills (pp. 7–8 in [33], and [34–38]).

2 High Performance Work and Self-Management

The idea of management has always been ready to serve the founders, owners and shareholders. Early years of management science witnessed the search for the better use of division of labor. Efficiency was the great cause, and practitioners and a limited number of scholars were involved in interminable debates over many concepts—e.g. *authority*, *order*, *hierarchy*, *centralization*, *and command chain*. They were, in matter and in manner, big words for a small human being. Eventually, along with the modern management era that began to shape satisfactorily during the post-war years corresponding to 1950s, managers of organizations either vast or tiny have been interested in incorporating the human factor into operations as a whole. This initiative took roots from the intention to consider the individual human as a whole—a whole body of biological and physiological attributes as well as talents, skills, level of mental acuteness, feelings, emotions and idiosyncrasies isolating him or her from other individuals. Company founders, owners and shareholders (and occasionally top managers) might seem to have been preoccupied due to two specific questions in their minds—two questions of which could never be completely disposed: *How can maximize the profit? And, will the company exist without cease?* These two basic questions actually reproduce multiple questions that need to get answered. In our case, we will demand a question based on the human factor: *Does high efficiency necessarily mean high quality and high performance?* Yet, this seems another challenging question—certainly not intact but looks like a doubtful bill.

2.1 Semler's Way: Reorganizing for Self-Management

Let's imagine a company with no organization chart, and no job titles. Neither ten year plans nor mission and vision statements exist. No dress code or no come and go times, no written rules... This bold idea of a company seems rather fictional as it conflicts with many topics in the field of management that are taught at school; however it is as true as life itself. Most of us—if interested in management theories or reading business books—might already be aware of the captivating story of Semco.

Having implemented the idea of *extremely* participative management into an enjoyable work life, Ricardo Semler has transformed Semco—a Brazilian manufacturing company of industrial machines from Sao Paulo, which was previously known as a giant entity where autocratic management and traditional methods were in practice—from top to down in this context. He published his experiences and the narrative of how he built such a culture in his company in a well-written book, *Maverick*, offering its readers a pleasant reading. The book was translated into thirty-six languages and sold millions of copies worldwide. He has also gained a global recognition. This spectacular success is convincing enough to be accredited as evidence to the influence he made in the minds of an international audience. Putting theory into practice is always a challenging process, and Semler has been considerably successful in doing it [39].

In his seminal book published in 1960, *The Human Side of Enterprise*, Douglas McGregor suggested an alternative approach the way managers approached their subordinates. McGregor's theory was offering a simple but rich insight to readers—trusting employees was the key to a better management of an organization. Trust is so crucial as a driving factor in social relationships that even in a bureaucratic setting, employees would feel themselves as part of this interlocked environment and show devotion to organization's objectives—particularly if they were treated as adults, as mature individuals [39]. Furthermore, placing confidence in employees to put their creativity and skills into practice and to carry out important decisions without hesitation or interference would serve to the advantage of the whole organization. The 'human nature' was the underlying argument in McGregor's debate and by nature, humans needed not 'carrots and sticks' type of incentives to work but—as highly appraised by Semler himself—*self-administration*, *self-governance*, and *self-control* (see particularly p. 58 in [40]).

The core idea of Semler's approach was to deliver employees more incentive in order to attract them to taking part in important decisions. The subsequent step was encouraging them to take part in self-managed teams—the reason to follow these steps was apparent: Given the opportunity to contribute to making important decisions, individuals would be more motivated, and in return, would make better decisions. As he illustrated in *Maverick* [41] the route to self-management was a rocky one, and the methodology used in walking or driving this rocky route was not soft as anticipated, just as Semler's first action had been firing nearly sixty per cent of the company's top managers in a single day! His argument—if not called a defence statement—was as simple as his core idea: Changing the mind-set of *conventional* managers was almost impossible. In fact, conventional manager was tantamount to managers accustomed to work ‘in his father’s way’. Semler’s words in *Maverick* provide a clear vision of his rightful argument:

> Semco appeared highly organized and well disciplined and we still could not get our people to perform as we wanted or be happy with their jobs. (p. 62 in [41])
>
> What if we could strip away all the artificial nonsense, all the managerial mumbo jumbo? What if we could run the business in a simpler, more natural way? (p. 63 in [41])

Discharging managers is indeed a courageous action. Once a final decision is made and arguments are settled, it may be as easy as tripping a few words of an explanation and some good luck wishes off the tongue. However, convincing and persuading employees for a change—in Semco’s case, for a drastic overhaul—is definitely one heavy duty for a boss. Semler’s heavy duty involved turning the company upside down and inside out. An unorthodox approach at that time against management theories was the basis of new practices. Nine layers of management were eliminated and an unprecedented type of corporate democracy was created. Since it seemed almost impossible to anticipate from an employee to transform immediately into an enthusiastic corporate individual who happily participates in making important decisions and take responsibilities as a member of a self-managed team, Semler decided to implement his plan gradually. He took initial steps by asking employees to generate solutions for small-scale problems like the improvement of the cafeteria food, or replacing overall colors with better ones. These baby steps helped employees realize the true adult-feeling essence of making decisions on their own issues. Many steps in quick succession empowered employees without the need for repulsive and annoying old policies and heavy bureaucracies that were created once upon a time. Below is the list of some of these revolutionary changes [41–43]:

- The traditional organizational chart was removed.
- Employees were given the freedom to set their own working hours and the responsibility to record them.
- Employees were encouraged to learn more about tasks other than their own, and to make suggestions for job improvements or departmental changes.
- Employees commenced to assess their supervisors’ efforts and performances and every six months, assessment results were announced straight out in the company.
- Employees were provided with the voice in recruitment affairs in their own departments and the right to vote for candidates.
- Rules and regulations that seemed useless, uncritical and indirect with job’s requirements like dress codes or reserved parking spaces were abolished.
- The workplace belonged to the employees—from painting, furnishing and flower planting to redesigning the layouts, employees acquired the right to make changes in their own working environment.
- Doors were declared open to unions for inspection.
- Corporate staff was eighty per cent less than any other comparable company.

As for Semler, employees were let to do whatever they wanted. In fact, Semco did not employ any performance analysts; there were only people who ‘sell or make’;

and this was not the outcome of the existence of two highly distinguished functional departments as one for marketing and one for production. In Semler's words:

> Marketing is everybody's problem. Everybody knows the price of the product. Everybody knows the cost. Everybody has the monthly balance sheet that says exactly what each of them makes, how much bronze is costing us, how much overtime we paid, all of it. And the employees know that 23 % of the after-tax profit is theirs. (p. 84 in [41])

Dumaine referred to the introduction of self-managed teams, or their variations, as 'the productivity breakthrough of the 1990s' with some clear and convincing evidence. Semler's implication on Semco's survival during times of meager economic circumstances, thanks to the company's 'increased flexibility as the result of the efforts of the self-managing employees' is one example. Additionally, Procter & Gamble reported 'between 30 to 40 % higher productivity' in the plants that utilized self-managed teams than in plants that did not. Food manufacturing companies General Mills and Johnsonville also put similar claims of higher productivity thanks to their self-managed teams [44–47]. As stated by Semler, 'it is up to them (employees) to see the connection between productivity and profit, and to act on it.' [48].

2.2 *Google's Way: Creating a Platform for Self-Management*

> So wouldn't you think that some of these computer companies would say, 'Apple does well because it has beautiful products'? So how much could the most expensive industrial designer in the world cost? You know, a million dollars a year in salary? Two million dollars a year? Five million dollars? That's what makes these products so beautiful. That level of commitment and dedication to the thing to make it the best thing possible.—Guy Kawasaki (author and former Apple Fellow) [49].

What is it that makes a great company more privileged than a merely good one? We have just witnessed that Semler was already aware of this question and the key to its answer even before taking the initiative and hitting a new trail. This time we will try to explore dynamics that emerge from within the organization—an organization that focuses on its social assets and sails on the waters of the blue ocean originated by its trustworthy individuals who dedicated themselves to creating and sharing knowledge.

Excellent companies thrive to find, recruit and retain the best people; and it is neither an exaggeration nor a myth that in certain cases, a company may find itself competing against other companies on the grounds of a human-resource-based battlefield. As put forward by Brown and Hesketh, 'where brand recognition is related to being a market leader, the best companies want to be seen to be recruiting the best talent, as these are mutually reinforcing.' (p. 85 in [50]) Besides, Smith and Kelly approach within a strategic human resource management framework embedded in the knowledge economy context, which we can comfortably observe in today's circumstances:

> … future economic and strategic advantage will rest with the organizations that can most effectively attract, develop and retain a diverse group of the best and the brightest human talent in the market place [51, 52].

At this point, one may immediately ask these obvious questions: Who are these *best people*? What signifies the *best talent*? A simple answer might be that best people are 'very intelligent people holding highest degrees and who have extensive knowledge, substantial experience and an exceptional background.' Once recruited, these *best* people, or calling it another way, these *great stars* are usually expected to bring miracles into their organizations and show outstanding performances that energize the work place environment. However, whether that great star is to create its own 'solar system' energizing planets and satellites that surround it and whether to fulfill such expectations usually remain as open-ended—if not rhetorical—questions. A company may hire its industry's greatest star at any cost; nevermore, incurring this risk does not always guarantee that the great star's social skills are compatible with organization's internal environment. Besides, research show evidence that these great stars are also risk bearers, which can potentially turn into comets. Moving from one company to another may wear away their potential and their exceptional performance usually drops drastically. A justification to this argument may be that the star's performance is predominantly dependent on the resources and the culture of the company rather than his or her capabilities. As suggested by Groysberg and colleagues in 2004, stars' success depends on such company-specific factors as 'resources and capabilities', 'systems and processes', 'leadership', 'internal networks', 'training', and 'teams' [53]. Not claiming that Kawasaki has meant this; but within this context, we find his words (quoted here a few paragraphs before) reminiscent and inspiring.

Companies with clear vision and managerial conscientiousness are able to hire and retain best talents through psychological contracts, whereas the structure of an employment relationship between the *great star* and the company is mutually shaped. This initiative can be extended into more specific contracts, *idiosyncratic deals*—special agreements that individuals negotiate with their employees (e.g. [54–57]). The content of such agreements and whether they generate dynamism in the organization is another debate; however, why building such a *loose and tight bond* between the employer and an employee who is expected to be a precious part of a high performance work system remains within our scope. Owing much to the vast literature on knowledge management and networks, cases on companies associated with the digital age usually give fruitful insights.

The case of Google sets a good example to the *best talent* culture as the world's leading Internet company hires exceptionally talented people and retains them in an ultimately mind-blowing workplace environment—usually envied by the best talented people who work in other companies. Being committed to having the most entrepreneurial environments in the world, Google aroused much interest in the public and gained image of a popular company for which a considerable number of MBA graduates wanted to work [58]. The company's way of attracting talents has even come up for a movie, *The Internship*, which exhibits the essence of their culture, referred to as *Googliness*, and people experiencing it as *Googley*. Well-known for its new product development strategies and capabilities for entering new business areas; Google is a good example to a company, where high performance work system is not created by a magic-wand-waving and spell-casting boss or manager but by constructing a platform—a platform for a high performance work system to *emerge*—where diverse talents could come together and get connected by means of social relationships and digital technology—the unique workplace environment, the popularity of products, the brand and everything else is only the reflection of it. Google's website states the company's culture with these words:

> It's really the people that make Google the kind of company it is. We hire people who are smart and determined, and we favor ability over experience. Although *Googlers* [59] share common goals and visions for the company, we hail from all walks of life and speak dozens of languages, reflecting the global audience that we serve [60].

Google also displays on a separate web page a list called *Ten Things Google has Found to be True*, which was prepared when the company was only a few years old. In point of fact, listed items reveal some of the fundamentals that underlie the company's latent but efficient high performance work policy from providing the best user experience possible to designing time-efficient, location-free and easy-to-use products. Particularly for one item, the statement goes;

> … work should be challenging, and the challenge should be fun. We believe that great, creative things are more likely to happen with the right company culture–and that doesn't just mean lava lamps and rubber balls. There is an emphasis on team achievements and pride in individual accomplishments that contribute to our overall success. [61]

With a semantics approach, the final item on the list seems almost similar to '*citius, altius, fortius*', the Olympic motto we have discussed earlier: 'Great just isn't good enough' [60]. A first impression about this item might be that Google supports performance based on a conventional type of competition:

> Do something, then check others and compare, and try to do something much better than others did!

However, the company explicitly describe on its webpage that the idea behind the item is about innovation and iteration:

> Do something, and if it works well then *improve upon it in unexpected ways*. (italics from [61])

In *The Future of Management*, authors Hamel and Breen compared the way Google was organized to the Internet itself. The determining factors of their argument were that both organizations were 'highly democratic', 'tightly connected', and 'flat'. Authority was disliked and a telling-people-what-to-do type of management practice was avoided since the company's inception. In Hamel and Breen's words, 'To Google's engineers, *Question Authority* is not an anarchist's bumper sticker, it's an innovator's imperative' (p. 109 in [62]). This explains satisfactorily why high performance work, according to Google culture, could be associated with efforts based on innovation and iteration rather than conventional competition.

Owing to perfectly capable and highly motivated employees already sharing a common vision, *Googlers* would never need to be micromanaged. Even so, leadership is appreciated in a company of hundreds of small self-managed teams. At the time when Hamel and Breen's book was published, about half of 10,000 employees at Google were working in small teams of three (on average). Per contra, large-scale projects like Gmail could employ thirty people or more, with partitioned into teams of three to four, each of which was assigned with one specific task. Each team had an '*über-tech* leader'; and these leaders rotated on teams, while engineers often worked on more than one project and were free to switch teams (pp. 111–112 in [62]). Despite all this haecceity, signs of crisis in the shape of 'increasing bureaucracy'; high performance and rapid growth of the company caused problems, particularly in sustaining the entrepreneurial culture; which caused highly talented employees to feel dissatisfaction at work and leave the company [58].

3 Conclusive Remarks

As debated earlier, self-managed teams may be regarded as essential parts of a high performance work system's infrastructure; and by all means, learning is *sine qua non*. According to Neck and colleagues, the developmental process of a self-managed team proceeds by a number of stages and the team continues to progress until its members finally arrive at the threshold where 'constructive synergistic thinking' is the key that opens the gate into 'the highest levels of performance' [63, 64]. In fact, as put forth by Senge in his invincible work in 1990, *The Fifth Discipline*, 'team-learning begins with dialogue, the capacity of members of a team to suspend assumptions and enter into genuine thinking together.' (p. 10 in [65]). Individuals at work want to think, share, collaborate, and enjoy this whole process all together. Self-managed teams emerge as an outcome of self-administration, self-governance, and self-control—all carried out in the possession of talented people selected with ensured cultural fit and diversity. Ricardo Semler was ready but Semco was not. He had to take the initiative to transform the company into a high performance organization. In the case of Google, the organization was ready. Google's employees were already creative, well-trained and self-confident; there was no need for micromanagement. The high performance organization emerged through their interactions and activities. Both companies had very different management styles; however both did the same thing.

On top of all this, forming up self-managed teams to achieve highest performance possible has always been a managerial concern and has never been easy. A specific management style to overcome this problem is not possible since the nature of self-managed teams itself is highly situational. Depending on countless dynamic factors—from human related problems to business environment conditions—managers as well as employees should assume the responsibility of forming up and carrying on the high performance work system.

References

1. Kling, J. (1995). High performance work systems and firm performance. *Monthly Laboratory Review, 118*, 29.
2. Zirger, B. J., & Maidique, M. A. (1990). A model of new product development: An empirical test. *Management Science, 36*(7), 867–883.

3. Clark, K. B., & Fujimoto, T. (1990). The power of product integrity. *Harvard Business Review, 68*(6), 107.
4. Angell, D. F., & Heslop, B. (1994). The internet business companion: Growing your business in the electronic age. Boston: Addison-Wesley Longman Publishing Co., Inc.
5. Van der Spek, R., & Spijkervet, A. (1997). Knowledge management: Dealing intelligently with knowledge. *Knowledge Management and its Integrative Elements*, 31–59.
6. Wiig, K. M. (1997). Knowledge management: An introduction and perspective. *Journal of Knowledge Management, 1*(1), 6–14.
7. Quintas, P., Lefrere, P., & Jones, G. (1997). Knowledge management: A strategic agenda. *Long Range Planning, 30*(3), 385–391.
8. Ulrich, D. (1998). Intellectual capital = competence × commitment. *Sloan Management Review, 39*(2), 15–26.
9. Baldwin, T. T., Danielson, C., & Wiggenhorn, W. (1997). The evolution of learning strategies in organizations: From employee development to business redefinition. *The Academy of Management Executive, 11*(4), 47–58.
10. Gollac, M., & Volkoff, S. (1996). Citius, altius, fortius. L'intensification du travail. *Actes de la recherche en sciences sociales, 14*, 54–67.
11. MacAloon, J. J. (1981). *This great symbol: Pierre de Coubertin and the origins of the modern Olympic Games*. Chicago: University of Chicago Press.
12. Peters, T. J., & Waterman, R. H. (2004). In search of excellence: Lessons from America's best-run companies. HarperCollins.
13. Saunders, J., & Wong, V. (1985). In search of excellence in the UK. *Journal of Marketing Management, 1*(1–2), 119–137.
14. Neely, A. (1999). The performance measurement revolution: Why now and what next? *International Journal of Operations and Production Management, 19*(2), 205–228.
15. Hitt, M. A., & Ireland, R. D. (1987). Peters and Waterman revisited: The unended quest for excellence. *The Academy of Management Executive, 1*(2), 91–98.
16. Bradford, D. L., & Cohen, A. R. (1997). *Managing for excellence: The guide to developing high performance in contemporary organizations*. New York: John Wiley & Sons.
17. Ouchi, W. (1981). Theory Z: How American business can meet the Japanese challenge. *Business Horizons, 24*(6), 82–83.
18. Pascale, R. T., & Athos, A. G. (1981). The art of Japanese management. *Business Horizons, 24*(6), 83–85.
19. Deal, T. E., & Kennedy, A. A. (1982). Corporate cultures: The rites and rituals of corporate life. Cambridge: Da Capo Press.
20. Kanter, R. M. (1983). Change masters. New York: Simon and Schuster Moss.
21. Ramsay, H., Scholarios, D., & Harley, B. (2000). Employees and High-Performance Work Systems: Testing inside the Black Box. *British Journal of industrial relations, 38*(4), 501–531.
22. Barney, J. B. (1995). Looking inside for competitive advantage. *The Academy of Management Executive, 9*(4), 49–61.
23. Huselid, M. A. (1995). The impact of human resource management practices on turnover, productivity, and corporate financial performance. *Academy of Management Journal, 38*(3), 635–672.
24. Harry, B. (1974). Labor and monopoly capital: The degradation of work in the twentieth century. New York: NYU Press.
25. Wood, S. (1999). Getting the measure of the transformed high-performance organization. *British Journal of Industrial Relations, 37*(3), 391–417.
26. Amin, A. (2000). Post-Fordism: models, fantasies and phantoms of transition. In A. Amin (ed.), *Post-Fordism: A reader* (pp. 1–40). Hoboken: Blackwell Publishers.
27. Aglietta, M. (1979). *A theory of capitalist regulation*. London: NLB.
28. Palloix, C. (1976). The labour process: from Fordism to neo-Fordism. The Labour Process and Class Strategies (London: Stage 1): pp. 46–67.
29. Evans, J. R., & Lindsay, W. M. (2005). The management and control of quality. Stamford: South-western Cengage Learning.
30. Kirby, J. (2005). Toward a theory of high performance. *Harvard Business Review, 83*(7), 30.

31. Gephardt, M. A., & Van Buren, M. E. (1996). Building Synergy: The Power of high performance work systems. *Training and Development, 50*(10), 21–36.
32. Kirkman, B. L., Lowe, K. B., & Young, D. P. (1999). High-performance work organizations: Definitions, practices, and an annotated bibliography. Center for Creative Leadership.
33. Appelbaum, E. (Ed.) 2000. Manufacturing advantage: Why high-performance work systems pay off. Ithaca: Cornell University Press,
34. Messersmith, J. G., Patel, P. C., Lepak, D. P., & Gould-Williams, J. S. (2011). Unlocking the black box: Exploring the link between high-performance work systems and performance. *Journal of Applied Psychology, 96*(6), 1105.
35. Ramstad, E. (2009). Promoting performance and the quality of working life simultaneously. *International journal of productivity and performance management, 58*(5), 423–436.
36. Birdi, K., Clegg, C., Patterson, M., Robinson, A., Stride, C. B., Wall, T. D., et al. (2008). The impact of human resource and operational management practices on company productivity: A longitudinal study. *Personnel Psychology, 61*(3), 467–501.
37. Whitfield, K. (2000). High-performance workplaces, training, and the distribution of skills. *Industrial Relations: A Journal of Economy and Society, 39*(1), 1–25.
38. Galang, M. C. (1999). Stakeholders in high-performance work systems. *International Journal of Human Resource Management, 10*(2), 287–305.
39. Fisher, L. M. (2005). Ricardo Semler won't take control. *Strategy and Business, 41*, 78.
40. McGregor, D. (2006). *The human side of enterprise-Annotated Edition*. Noida: Tata McGraw-Hill Education.
41. Semler, R. (2001). *Maverick: The success story behind the world's most unusual workplace*. New York: Random House.
42. Vanderburg, D. (2004). The story of Semco: The company that humanized work. *Bulletin of Science Technology and Society, 24*(5), 430–434.
43. Wieners, B. (2004). Ricardo Semler: Set them free. CIO Insight.
44. De Waele, M. (1994). Self-Management: Road to a productive, healthy and self-sustaining organizational Milieu? *Human Systems Management, 13*(3), 153–164.
45. Dumaine, C. A. (1990). Who needs a boss? Fortune *121*, 10.
46. Semler, R. (1994). Why my former employees still work for me. *Harvard Business Review, 72*(1), 64–71.
47. Waterman, R. H, Jr. (1995). What America does right: Learning from companies that put people first. *Business Horizons, 38*(1), 90–91.
48. Semler, R. (1989). Managing without managers. *Harvard Business Review, 67*(5), 76–84.
49. Excerpt from Welcome to Macintosh (aired at 07:30 pm on January 31, 2010). Transcript retrieved from http://livedash.ark.com/transcript/welcome_to_macintosh/5406/CNBC/Sunday_January_31_2010/181790 on Jan 11, 2014.
50. Brown, P., & Hesketh, A. (2004). *The mismanagement of talent: Employability and jobs in the knowledge economy*. Oxford: Oxford University Press.
51. Smith, A. F., & Kelly, T. (1997). Human capital in the digital economy. *The organization of the future*. (199–212) San Francisco, CA: Jossey-Bass Publishers.
52. Thite, M. (2004). Strategic positioning of HRM in knowledge-based organizations. *Learning Organization, The, 11*(1), 28–44.
53. Groysberg, B., Nanda, A., & Nohria, N. (2004). The risky business of hiring stars. *Harvard Business Review, 82*(5), 92–101.
54. Berber, A. (2014). Psychological contracts in the age of social networks. *In Human Resource Management and Technological Challenges* (pp. 23–42). Springer International Publishing.
55. Rousseau, D. M. (2005). I-deals: Idiosyncratic deals employees bargain for themselves. Armonk: ME Sharpe.
56. Anand, S., Vidyarthi, P. R., Liden, R. C., & Rousseau, D. M. (2010). Good citizens in poor-quality relationships: Idiosyncratic deals as a substitute for relationship quality. *Academy of Management Journal, 53*(5), 970–988.
57. Rousseau, D. (1995). Psychological contracts in organizations: Understanding written and unwritten agreements. Sage.

58. Finkle, T. A. (2012). Corporate entrepreneurship and innovation in silicon valley: The case of google, Inc. *Entrepreneurship Theory and Practice, 36*(4), 863–884.
59. Googlers are employees working at Google.
60. http://www.google.com/about/company/facts/culture. Retrieved on Jan 14, 2014.
61. http://www.google.com/about/company/philosophy. Retrieved on Jan 15, 2014.
62. Hamel, G., & Breen, B. (2007). The future of management. Harvard Business Press.
63. Power, J., & Waddell, D. (2004). The link between self-managed work teams and learning organisations using performance indicators. *Learning Organization The, 11*(3), 244–259.
64. Neck, C. P., Connerley, M. L., & Manz, C. C. (1997). Toward a continuum of self-managing team development. *Advances in interdisciplinary studies of work teams, 4*, 193–216.
65. Senge, P. M. (2006) The fifth discipline: The art and practice of the learning organization. New York: Random House Digital, Inc.

The Role of the Line Manager in HRM-Performance Research

Brian Harney and Kenneth Cafferkey

Abstract This chapter will review the evolution of research examining the HRM-Performance relationship. In particular it will focus on the gradual incorporation of the line manager as a critical agent in this relationship. The chapter then presents a conceptual schema which facilitates in better delineating the role and impact of line manager interventions along various units of analysis. The chapter concludes with implications for practice and future research trajectories.

1 Introduction

There is growing recognition that the management of people is of strategic significance to success [1]. A burgeoning stream of international evidence over the last two decades has highlighted the impact that Human Resource Management (HRM) can have in shaping both organisational performance [2] and employee outcomes [3]. Yet while a consensus has gradually emerged about the content of HRM practices [4], what is still underdeveloped is our understanding about the precise mechanisms through which HRM's impact takes effect. This chapter will engage with this issue by focusing on HRM from a process perspective [5], paying particular attention to the role of the line manager as a central actor in the HRM-performance relationship. The chapter begins by charting the evolution of HRM-Performance research noting the gradual incorporation multiple stakeholder

B. Harney (✉)
Dublin City University Business School, Dublin City University, Dublin 9, Ireland
e-mail: brian.harney@dcu.ie

K. Cafferkey
Graduate School of Business, Universiti Tun Abdul Razak, Tun Razak City Campus, Block C&D, Capital Square, No. 8, Jalan Munshi Abdullah, 50100 Kuala Lumpur, Malaysia

C. Machado and J. P. Davim (eds.), *Work Organization and Human Resource Management*, Management and Industrial Engineering,
DOI: 10.1007/978-3-319-06376-8_3,

perspectives. It then goes on to highlight the significance of line management interventions and maps the various impact and roles of line managers in a holistic fashion. The chapter concludes by outlining implications for future research.

2 Evolution in Understanding the HRM-Performance Relationship

The importance of HRM came to the fore in the context of intensive global competition, most notably from Japan, and the emergence of more people-intensive service and knowledge-based industries. The logic of the resource based view provided a theoretical underpinning to explore the role of internal organisational processes such as HRM in generating competitive advantage [1]. HRM's impact has been subsequently demonstrated empirically by numerous studies which find that sophisticated HRM practices are positively related to organisational performance [6]. Some of the labels that have been attached to HRM highlight the nature of this linkage, for example best practice HRM, high performance work systems (HPWS) and high commitment management (HCM) [7]. In the case of HPWS, these have been defined as 'systems of human resource practices designed to enhance employees' skills, commitment and productivity in such a way that employees become a source of competitive advantage' [8]. In this definition and others there is a widely held view that SHRM impacts upon organisational performance through practices that promote and nourish employee autonomy, skills and discretionary behaviours. HRM practices typically included in these studies include sophisticated selection, appraisal, training, teamwork, communication, job design, empowerment, participation, performance-related pay/promotion, harmonization, and employment security. There is also a particular emphasis on how these practices work together as a mutually reinforcing bundle [2].

More recently it has been appreciated that while HRM practices may be necessary for enhanced performance, on their own they are hardly sufficient [9]. Research has in turn moved beyond simple counts of practices offering 'growing sophistication' into the mechanisms by which HRM may impact performance. According to Sanders and Frenkel [5] this represents a move from best practice HRM to a process perspective. Critical here has been a distinction between intended HR practices as designed by HR departments, implemented HR practices as those practices actually enacted by line managers, and perceived practices as those practices experienced by employees [10]. Rendering the causal chain of HRM's impact more transparent in this manner has opened up numerous research streams including those informed by social exchange theory, the Ability-Motivation-Opportunity framework in addition to research attentive to multiple actors and multifaceted measures of performance [11]. Notably, because of their role as the 'touch point between the intended practice and the employee as recipient' [12] line managers have emerged as critical components in the HRM-performance chain.

3 Line Managers as Critical Intermediaries

The significance of line managers as critical intermediaries in the HRM-performance relationship has slowly been recognised. Conceptually work by McDermott and colleagues [13] illustrates how differing leadership styles need to be aligned with strategy and are also likely to influence the nature of the psychological contract held by employees. Empirically work by Purcell and colleagues [14] has illustrated how line managers are 'a crucial variable in how policies, practices and values are brought to life'. Redman and Snape [15] found that line managers have a greater influence over their subordinates than top or middle management. Focusing on research orchestration, Chadwick et al. [16] find that line managers are vital in ensuring consistency between top management intent and organisational activity. They note how 'senior leaders give supervisors the responsibility and authority to earn the enthusiasm, energy, and creativity that signal deep employee engagement'. In this guise line managers act as key gate keepers ensuring a clear and consistent message is provided to employees in terms of HPWS policy [17]. Recent practitioner reports also recognize the role that line managers play in fostering 'a climate that supports high performance and engagement' while also constituting 'the single biggest influence on an employee's development' [18]. A frequent assertion that reinforces this point is the notion that employees do not leave organisations, rather they leave managers.

Yet while the role of line managers may is increasingly acknowledged [12, 19] typically lacking is a more systematic assessment and conceptualisation of the potential role and impact of line managers in shaping and sustaining the relationship between HRM and performance [20]. Purcell et al. [14] addressed the importance of line management as catalysts for the performance of their team members, however employee based research is still in its infancy so that in-depth explorations of the influence of line managers is still underdeveloped. Addressing this deficiency is important as ultimately 'the HR practices that employees perceive and experience will be heavily influenced by the quality of their relationship with their direct manager' [21]. In order to advance understanding the next section of this chapter presents a schema to capture the various roles and levels of influence that line managers may exert in the HRM-performance relationship.

4 Advancing Understanding of Line Managers in the HRM-Performance Relationship

Figure 1 depicts line manager roles and activities at both organisational and individual units of analysis. It also distinguishes between interventions and theories which might be seen as more collective in orientation as opposed to those focusing on the individual. Each domain is therefore associated with varying theoretical perspectives and tensions with differing implications for research.

Quadrant 1 is associated with the more traditional emphasis in HRM-performance studies whereby the focus is on strategic integration and organisational

Level of analysis / Orientation	Collective	Individual
Organisation	1) Strategic HRM High Performance Work Systems Intended HRM	2) Role Theory Identity Implemented HRM
Employee	3) Climate Social exchange theory Implemented HRM/ Experienced HRM	4) Local commitment Psychological contract Experienced HRM

Fig. 1 Categorising line management interventions

level performance. Research here is typically of the intended HR variation [10] with a focus on organisational level policies and their likely impact on organisational performance. Research on line managers from this perspective has focused on practices likely to impact their ability, motivation and opportunity to deliver organisational objectives. Indeed, early research either implicitly or explicitly drew upon the logic of the AMO framework, which has become something of a referent for HR studies [22]. In terms of organisational outcomes research by Roche and Teague [23] demonstrates that proactive line manager engagement in conflict management is positively related to organisational outcomes and that this relationship is further enhanced by the presence of HPWS types of HR practices. At the employee level, studies have recognised the importance of supervisory support as a key variable determining organisational commitment [24]. It is clear from research associated with this quadrant that both policies for line managers and activities by line managers can exert significant influence at an organisational level. Research questions here consider the nature of HRM practices in place and the nature of HRM practices that line managers are subject to, with a particular focus around practices which are directly intended to enhance the ability of line manager's to perform their tasks, the recognition and motivation they are prescribed to receive for so doing, and finally the organisational infrastructures in place that is designed to afford line mangers an opportunity to conduct HR activities in a desired and valuable fashion [14].

Quadrant 2 retains an organisational level focus but explores its implications for individual relationships. Here we might consider role theory and possible role conflicts for line managers in trying to reconcile their role as resource orchestrator cascading top management intent on one hand, while also being open and accommodating of the realities of employee experiences on the other [25]. This suggests that the traditional tensions and ambiguities surrounding identity and influence experienced by personnel/HR professions [26] hold relevance as migrated to the role of the line manager. Line management role ambiguity and conflict have been particularly noted in the realm of performance appraisal in part due to lack of line management training and buy-in manifest in poor commitment to conducting HR activities [27]. Research by Geare and colleagues [11] found a clear distinction between unitarist ideological intent expressed by management and more pluralist orientations expressed by employees. Similar to HR managers, line managers are likely to be caught in the midst of this dynamic on a day to day basis. This will have implications in terms of line manager's identity and also their willingness to engage in HR activities. Research streams associated with this quadrant examine the differing expectations of line managers and HR managers concerning their roles and status and what HR activities they should conduct [28]. Research also examines the individual tensions confronted by line managers as they strive to realise their HR ambitions while also directly managing their workforce. Research on identity is likely to be particularly relevant here. As noted by Thomas and Davies [29] 'identity can be usefully conceptualized as a dynamic struggle between various subject positions within discourses, vying for our attention and our active human agency. Thus individuals are situated in social contexts that both constrain and sustain identity'. Taking a micro-political approach leaves an analytical space to explore how line managers may subvert and resist attempts to alter their roles and to explore their response to attempts to directly incorporate them as an assumed part of an organisational level performance agenda.

Quadrant 3 examines collective orientation but focuses on employee level outcomes. Here the emphasis is on those practices actually implemented by line managers and the collective employee experiences of such practices as captured through social exchange theory and manifest in organisational climate [30]. HR systems have been suggested as an important antecedent to organizational climate. The mediating variable between the HR system and employee experiences is the activity and behaviour of line management. The role of the line manager in fostering a collective culture becomes critical in shaping employee shared perceptions of HR. Theoretical perspectives here bring together both line manager and employee perspectives to offer accounts of the social climate within the firm and its role in bridging intended and enacted HR practice [3, 30]. It is argued that such a social system or the social context of work [31] is fundamental in eliciting discretionary effort or organisational citizenship behaviours whereby employees perceive HR practices through their enactment by line managers [32]. The social system, in turn, facilitates greater access to relevant information, reduces the need for control and also allows for collective action [33]. Although HR practices are seen to operate as macro concepts the effect of such HR initiatives are processed and realised at the individual and group level [3]. Ferris et al. [31] suggest a 'social interaction processes' whereby individual employees determine organisational reality and meaning from relationships, norms and values within the group, in an intra- organisational environment. Takeuchi et al. [3] supports this view suggesting that shared meaning occurs due to 'ambient social and structural stimuli operating on all members of the same unit (e.g. exposure to similar norms, leadership, and HR practices)'. Bommer et al. [34] state that there is often an omission in the relationship between an individual work environment and positive organizational citizenship behaviours where 'many forms of human behaviour are influenced by the attitudes and behaviours of the immediate community'. Becker [35] suggests that individuals identify with, and adopt, attitudes and behaviours that they consider to be associated with a satisfying relationship with other individuals within a group. These processes then become internalised as employee actions become congruent with those of the group. This view is remarkably different from the conventional propositions of HPWS that suggests commitment and satisfaction are directed toward the organisation as a result of HR interventions.

Aligned with this perspective research by Arrowsmith and colleagues [36] has expanded the ability-motivation-opportunity (AMO) framework to incorporate an additional dimension of 'direction'. This is understood to include the elaboration and communication of expectations, relevant performance measures, feedback, coaching and understanding and application of consequences [36]. Arguably much of these activities are associated with line management interventions in enacting HRM practice. Line managers are critical in creating the work environment and the informal side of the organisations, ensuring operational efficiency and the implementation of organisational objectives, and therefore are deemed decisive in determining attitudes and behaviours of employees. More process based perspectives have moved to appreciate this insight, focusing less on the intended practices associated with traditional research (quadrant 1) and moving to incorporate multiple stakeholders. It follows that research in this quadrant tends to offer more multi-level insights and incorporates multiple respondents. Frequently rich insights about the process of enacting HR come from more in-depth qualitative approaches.

Quadrant 4 examines the individual relationship between line manager and employees. Increasingly there is recognition that while individuals may not be committed to their organization per se they may be highly committed to their department, profession, union or potentially to their line manager [15]. The role of leader–member exchange theory provides useful avenues to explain the reasoning behind why employees may see HR activities purely as supervisor initiatives [32]. Supervisory support has been shown to be an important influence on the implementation of HR practices. According to Bowen and Ostroff [37]: 'supervisors can serve as interpretive filters of HRM practices and can foster stronger relationship between HRM, climate, and performance than would exist individually'. Litwin and Stringer [38] noted that support and encouragement by one's immediate supervisor is of paramount importance for the arousal of achievement. Notably, Purcell and Hutchinson [32] found that employees viewed practices such as communication and involvement solely through their supervisor and not as organisational initiatives. As such it can be said that HPWS initiatives are viewed in relation to the social system based on relationships and therefore employees appeared oblivious to organisational policies as such policies were actually interpreted as supervisor actions in isolation. It follows that employee-supervisor relationships may therefore be the foundation on which the success of HRM interventions rest.

Harney and Jordan [19] provide case study evidence from a call centre in Ireland where, despite HRM being imposed by a dominant supplier, line managers' interventions ameliorated some of the negative aspects of work tasks. This included the introduction of practices which resonated with those prescribed by HPWS, including efforts at improving morale and creating a sense of involvement and a better atmosphere among telesales representatives. The result was that while employees exhibited the low commitment characteristic of the sector, they also demonstrated significant organisational citizenship behaviours manifest in discretionary efforts to go beyond what was required of the immediate task on hand. Rather than assuming that line management behaviour will automatically align with intentions [13] this approach allows for a possible divergence between what is intended of line management and what is actually enacted by line managers. Thus consistency may not necessarily be the final arbiter of successful performance as emergent leadership behaviour can be critical. Likewise, individual demographics including age, experience, gender and cultural background may shape the enactment and meaning attributed to interventions [39]. Research in this quadrant typically focuses on the interaction between line managers and employees, paying particular attention to micro-level variables and outcomes including foci of commitment, engagement and the nature of the psychological contract held by employees. For example, research has shown that concepts such as local commitment (i.e. commitment to supervisor) may find greater significance than organisational level commitment [16, 19].

5 Discussion and Summary

The critical role of line managers in shaping the HRM-performance relationship is something increasingly recognised by researchers and practitioners alike [18, 40]. Numerous studies demonstrate that line managers are the primary contract makers for employees [13]. This chapter presented a conceptual schema which captures the various roles and impact of line management interventions. This provides a holistic and systematic understanding of line management behaviour thereby illuminating the various ways line managerial interventions may facilitate in achieving 'implementation advantage' in the domain of HRM [5]. Greater understanding in this area is all the more important in light of emerging organisational forms and increased responsibility and accountability being devolved down to the line [32].

Exploration of line management interventions is timely for the HRM-Performance field as while the support of top management and introduction of sophisticated HRM practices are necessary for HRM performance impact, it is increasingly recognised that they are not sufficient. As recently noted by Chadwick and colleagues [16] 'although top management tends to set the overall direction and strategic emphasis of the firm, the responsibility for actually implementing strategy often falls to managers and employees lower down in the firm'. Line managers are therefore critical in terms of leveraging desired and appropriate organisational and especially employee outcomes [16]. Specifically, line managers create an infrastructure which bridges managerial intent and employee experiences [30]. This speaks directly to recent emphasis on the processes of how and why HRM influences performance, most notably to the process orientated inter-personal relationships within organizations [41]. To date these relationship have largely been neglected despite the fact that the process and quality of employee relationships can act as a critical determinants of performance [42].

By categorizing various line manager roles this chapter has illustrated the multiple means by which line managers act as a vital component in creating a social structure or system [32]. In a social system reciprocation is towards the group and supervisors, and it is this process of 'conscientiousness' which in turn further strengthens and reinforces the system [10]. This resonates with the group process literature [43] where values and beliefs are created at the group level. Group norms are more proximal to performance may therefore have a stronger influence than HR practices. Indeed such norms may actually obstruct or counteract the effect of HRM practices [10]. Consideration of line managers as presented in this chapter therefore offers a pathway to incorporate recent calls for more multi-level insights of the HRM-performance relationship including the relatively neglected roles of group and team level phenomenon.

6 Conclusion

Overall, our consideration of line managers links with a newly conceptualised version of HRM understood as incorporating 'signals from management to employee groups and individuals' [44]. Haggerty and Wright's [45] extension of the Bowen and Ostroff arguments lies in their suggestion that 'the strength of the HR system is leveraged more on the process and less on the content of HRM'. Allied to this Lau and Liden [46] suggest that interpersonal trust and reciprocation is directly influenced by a third party; the supervisor. They suggest that the supervisor as the official appraiser, reward distributer and mentor of employees shapes the social system in which employees operate, where they in effect 'set the tone for the relationships within the work group and influence the environment in which the work group is embedded'. Line mangers are clearly central to advancing our understanding of the dynamics of managing the employment relationship. Key challenges for researchers in this area include reaching a consensus on a precise definition of what actually constitutes line management while also mastering the complexities of multi-level research and multi-faceted dimensions of performance. The schema presented in this chapter is one attempt to weave together extant contributions. In so doing it opens up multiple research pathways for exploring the interaction of organisational or employee levels of analysis coupled with the collective or individual orientation. Exploring this schema will facilitate researchers in asking further important questions. The research trajectories that ensue will facilitate in empirically demonstrating the significance line manager interventions in a systematic fashion of direct benefit to both research and practice.

References

1. Barney, J. (1991). Firm resources and sustained competitive advantage. *Journal of Management, 17*(1), 99–120.
2. Huselid, M. A. (1995). The impact of human resource practices on turnover, productivity, and corporate financial performance. *Academy of Management Journal, 38*(3), 635–672.
3. Takeuchi, H., Chen, G., & Lepak, D. P. (2009). Looking through the looking glass of a social system: Cross level effects of HPWS on employees' attitudes. *Personnel Psychology, 62*, 1–29.
4. Posthuma, R., Campion, M., Masimova, M., & Campion, M. (2013). A high performance work practices taxonomy: Integrating the literature and directing future research. *Journal of Management, 39*(5), 1184–1220.
5. Sanders, K., & Frenkel, S. (2011). HR-line management relations: Characteristics and effects. *International Journal of Human Resource Management, 22*(8), 1611–1617.
6. Combs, J. G., Ketchen, D., Hall, A., & Liu, Y. (2006). Do high performance work practices matter? A meta-analysis of their effects on organisational performance. *Personnel Psychology, 59*, 501–528.
7. Wall, T., & Wood, S. (2005). The romance of human resource management and business performance, and the case for big science. *Human Relations, 58*(4), 429–462.
8. Datta, D., Guthrie, J., & Wright, P. (2005). Human resource management and labor productivity: Does industry matter? *Academy of Management Journal, 48*(1), 135–145. (p. 135).
9. Patel, P., Messersmith, J., & Lepak, D. (2013). Walking the tightrope: An assessment of the relationship between high-performance work systems and organizational ambidexterity. *Academy of Management Journal, 56*(5), 1420–1442.
10. Wright, P., & Nishii, L. (2013). Strategic HRM and organizational behavior: Integrating multiple levels of analysis. In D. Guest, J. Paauwe, & P. Wright (Eds.), *HRM and performance: Building the evidence base* (pp. 97–110). San Francisco, CA: Wiley.
11. Geare, A., Edgar, F., McAndrew, I., Harney, B., Cafferkey, K., & Dundon, T. (2014). Exploring the ideological undercurrents of HRM: Workplace values and beliefs in Ireland and New Zealand. International Journal of Human Resource Management, Forthcoming.
12. Brewster, C., Gollan, P., & Wright, P. (2013). Guest editor's note: Human resource management and line. *Human Resource Management, 52*(6), 829–838. (p. 831).
13. McDermott, A., Conway, E., Rousseau, D. M., & Flood, P. (2013). Promoting effective psychological contracts through leadership: The missing link between HR strategy and performance. *Human Resource Management, 52*(2), 289–310.
14. Purcell, J., Kinnie, N., Hutchinson, S., Rayton, B., & Swart, J. (2003). Understanding the people and performance link: Unlocking the black box. London: Chartered Institute of Personnel Development (p. 11).
15. Redman, T., & Snape, E. (2005). Unpacking commitment: Multiple loyalties and employee behaviour. *Journal of Management Studies, 42*(2), 301–328.
16. Chadwick, C., Super, J. F., & Kwon, K. (2014). Resource orchestration in practice: CEO emphasis on SHRM, commitment-based HR systems and firm performance. *Strategic Management Journal,*. doi:10.1002/smj.2217.
17. Boxall, P., & Purcell, J. (2008). Strategy and human resource management (2nd ed.). London: Palgrave MacMillan (p. 218).
18. Shanks, R., O'Neill, N., & O'Mahony, A. (2013). Closing the skills gap in Ireland: Employers at the heart of the solution, Accenture Institute for High Performance, Accenture. http://www.accenture.com/ie-en/Pages/insightsolving-skills-paradox-seven-ways-close-critical-skills-gaps.aspx
19. Harney, B., & Jordan, C. (2008). Unlocking the black box: Line managers and HRM performance in a call centre context. *International Journal of Productivity and Performance Management, 57*(4), 275–296.
20. Huselid, M. A. (2011). Celebrating 50 years: Looking back and looking forward: 50 years of human resource management. *Human Resource Management, 50*(3), 309–312.

21. Boxall, P., & Purcell, J. (2008). Strategy and human resource management (2nd ed.). London: Palgrave MacMillan (p. 220).
22. Jiang, K., Lepak, D., Hu, J., & Baer, J. (2012). How does human resource management influence organizational outcomes? A meta-analytic investigation of mediating mechanisms. *Academy of Management Journal, 55*(6), 1264–1294.
23. Roche, B., & Teague, P. (2012). Do conflict management systems matter? *Human Resource Management, 51*(2), 231–258.
24. Patterson, M., West, M., Shackleton, V., Dawson, J., Lawthorn, R., Maitlis, S., et al. (2005). Validating the organisational climate measures: Links to managerial practices, productivity and innovation. *Journal of Organisational Behaviour, 26*, 379–408.
25. Harney, B. (2014). Managerial roles. In P. Flood & Y. Freeney (Eds.), *Wiley encyclopedia of management* (3rd ed.). Hoboken: Wiley-Blackwell.
26. Caldwell, R. (2003). The changing roles of personnel managers: Old ambiguities, new uncertainties. *Journal of Management Studies, 40*(4), 983–1004.
27. Harris, L. (2001). Rewarding employee performance: Line manager's values, beliefs and perspectives. *International Journal of Human Resource Management, 12*(7), 1182–1192.
28. Whittaker, S., & Marchington, M. (2003). Devolving HR responsibility to the line: Threat, opportunity or partnership? *Employee Relations, 25*(3), 245–261.
29. Thomas, R., & Annette, D. (2011). Reassessing identity: The relevance of identity research for analysing employment relations. In P. Blyton, E. Heery & P. Turnbull (Eds.), *Reassessing the employment relationship* (pp. 147–170). London: Palgrave MacMillan (p. 148).
30. Bowen, D., & Ostroff, C. (2004). Understanding HRM-performance linkages: The role of the 'strength' of the HRM system. *Academy of Management Review, 29*(2), 202–222.
31. Ferris, G. R., Arthur, M., Berkson, H., Kaplan, D., Cook-Harrell, G., & Frink, D. (1998). Towards a social context theory of human resource management-organization effectiveness relationship. *Human Resource Management Review, 8*(3), 235–264.
32. Purcell, J., & Hutchinson, S. (2007). Front-line managers as agents in the HRM-performance causal chain: theory, analysis and evidence. *Human Resource Management Journal, 17*(1), 3–20.
33. Evans, W. R., & Davis, W. D. (2005). High-performance work systems and organizational performance: The mediating role of internal social structure. *Journal of Management, 31*(5), 758–775. (p. 759).
34. Bommer, W. H., Miles, E. W., & Grover, S. L. (2003). Does one good turn deserve another? Coworker influences on employee citizenship. *Journal of Organizational Behavior, 24*, 181–196. p. 182.
35. Becker, T. (1992). Foci and bases of commitment: are they distinctions worth making? *Academy of Management Journal, 35*(1), 232–244. (p. 133).
36. Arrowsmith, J., & Parker, J. (2013). The meaning of 'employee engagement' for the values and roles of the HRM function. *International Journal of Human Resource Management, 24*(14), 2692–2712.
37. Bowen, D., & Ostroff, C. (2004). Understanding HRM-performance linkages: The role of the 'strength' of the HRM system. *Academy of Management Review, 29*(2), 202–222. (p 215–216).
38. Litwin, G., & Stringer, R. (1968). *Motivation and organizational climate*. Boston: Harvard University. (p. 52).
39. Nishii, L., Lepak, D. P., & Schneider, B. (2008). Employee attributions of the 'why' of HR practices: Their effects on employee attitudes and behaviors, and customer satisfaction. *Personnel Psychology, 61*(3), 503–545.
40. Freeney, Y., & Fellenz, M. (2013). Work engagement, job design and the role of the social context at work: Exploring antecedents from a relational perspective. *Human Relations, 66*(11), 1427–1445.
41. Evans, W. R., & Davis, W. D. (2005). High-performance work systems and organizational performance: The mediating role of internal social structure. *Journal of Management, 31*(5), 758–775.
42. Collins, C., & Smith, K. G. (2006). Knowledge exchange and combination: The role of human resource practices in the performance of high-technology firms. *Academy of Management Journal, 49*(3), 544–560.

43. Kozlowski, S. W. J., & Bell, B. S. (2003). Work groups and teams in organizations. In W. Borman, R. Liden, & R. Klimoski (Eds.), *Handbook of psychology: Industrial and organizational psychology*: (Vol. 12, pp. 333–375). New York: Wiley.
44. Haggerty, J., & Wright, P. (2010). Strong situations and firm performance: A proposed reconceptualisation of the role of the HR function. In A. Wilkinson, N. Bacon, T. Redman & S. Snell (Eds.), *The sage handbook of human resource management*. London: Sage (p. 101).
45. Haggerty, J., & Wright, P. (2010). Strong situations and firm performance: A proposed reconceptualisation of the role of the HR function. In A. Wilkinson, N. Bacon, T. Redman & S. Snell (Eds.), *The sage handbook of human resource management*. London: Sage (p.106).
46. Lau, D., & Liden, R. (2008). Antecedents of co-worker trust: Leaders blessings. *Journal of Applied Psychology, 93*(5), 1130–1138.

The Impact of Strategic Human Resource Management Practices on Australian Dairy Farm Performance

Aman Ullah and Connie Zheng

Abstract The purpose of this Chapter was to assess the impact of human resource management (HRM) practices adopted by 205 Australian dairy farmers on their business performance. Factor analysis extracted ten factors out of sixty items that represented all possible HRM practices in the farming context. Ten HRM practice factors were subsequently entered into multiple regression analysis as explanatory variables; these were formal selection process, hiring procedure, training, performance reviews, communication process, career opportunities, risk management of occupational health and safety (OH&S), informal social interaction, compliance-based HR records and standard milking procedures. Dependent variables were financial outcome measured by percentage increase in farm profitability and net profit shared with employees; and herd health measured by percentage of cows treated for mastitis, culled and death rates due to poor health conditions. Four business strategies (i.e. cost reduction, product quality, technology innovation and people management) were used as control variables. Regression results show a clear link of HRM to farm performance, despite several formal HRM practices (i.e. validated hiring, selection and training) negatively related to dairy farms' financial outcome however positively related to herd health. The results also show that informal HRM practices (i.e. daily & peer performance review, communication and social interaction) played more significant roles in improving financial outcome; whilst HRM practices integrating with business strategies (i.e. product quality, technology innovation and people management) predicted better herd health in dairy. Implications of these results are discussed.

A. Ullah
Department of Livestock Economics and Business, University of Veterinary and Animal Sciences, Lahore, Pakistan

C. Zheng (✉)
Deakin Graduate School of Business, Faculty of Business and Law, Deakin University, 70 Elgar Road, Burwood, VIC 3125, Australia
e-mail: connie.zheng@deakin.edu.au

C. Machado and J. P. Davim (eds.), *Work Organization and Human Resource Management*, Management and Industrial Engineering,
DOI: 10.1007/978-3-319-06376-8_4,

1 Introduction

The literature examining the impact of human resource management (HRM) practices on firm performance has been accumulated, however largely focused on industries other than dairy, such as manufacturing [1], financial sector [2], and services [3]. A pioneering work by Stup et al. investigated the relationships between selected HRM practices and dairy farm performance based on a sample size of eighty Pennsylvania dairy farms [4]. It is found that those dairy farms that used the technical HRM practices (e.g. the use of continued training and standard operating procedures for feeding) performed much better in the areas of return on equity and somatic cell count than those that did not [4–6]. Yet, the evaluation of synergistic effects of both technical and strategic HRM practices on dairy farm performance is yet completed [4, 7, 8], even less so in the Australian context. This leaves much room for speculating what strategic value HRM could add to creating a high-performing workplace in the farming context—the key concern widely discussed in the field of strategic HRM [9] as well as in Australia's 'People in Dairy Program' [10].

Strategic HRM is defined as a process of linking organizational business strategies with internally coordinated HRM policies/practices and to externally changing context in order to achieve organizational performance objectives [11]. The essence is to develop organizational HRM policies/practices that would support the implementation of chosen business strategies and address the needs of various stakeholders [9]. In the context of dairy, the stakeholders consist of not just farmhands, employees or family members, but external bodies such as dairy industry association, government and rural and regional community [12].

Generic business strategies developed earlier by Michael Porter cover three distinctive strategies (i.e. cost leadership, quality and innovation) based on organizational market scale and competitive positioning [13]. In line with these competitive business strategies, Schuler and Jackson argued that organizations must develop corresponding set of HRM practices to help implement chosen business strategies [14]. In recent years, strategic labour management skills were identified as the most important skill to achieve successful expansion in the American dairy [15, 16]. Yet it is unclear what would be the distinctive set of HRM practices that dairy farm owners/managers would have developed to support their chosen business strategies (be it cost reduction, product quality, technology innovation etc.) in order to achieve better farm performance.

For the past decade, the Australian dairy industry, similar to those in Canada, New Zealand, and the USA, has not only confronted internal structural changes, but also been required to meet external legislative requirements especially in the areas of employment conditions and occupational health and safety (OHS). The structural changes are reflected in a reduced number of registered Australian dairy farms to 7,511 (2009/10) from nearly 22,000 in 1979–80; and increase in average herd size currently at 274 cows, up from 85 cows in 1980 [17]. The increases in average herd size demand for more paid employees, and changing managerial roles to focus more on strategic issues of farm operation and risk management of compliance versus commitment-based human resources [4, 15–20]. One of the key challenges for Australian dairy farmers was to meet the legislative requirements with new workplace relation laws as a result of frequent changing governments and ideology shifts between the Labour and Coalition parties.

In light of the above-mentioned context and potential managerial implications for developing a sustainable dairy industry, this study was set to examine the impact of strategic HRM on farm performance, using Australia as a case study.

2 Materials and Methods

Because of ambiguity about a synergistic set of HRM practices likely to be implemented in the dairy farming context, extensive literature review, a focus group discussion and ten face-to-face interviews with industry experts were conducted during 2009–2010 to gather sixty items of all possible HRM practices relating to recruitment and selection, training and career development, performance evaluation, open communication, employee socialisation, compliance-based HR records, OHS and standard operating procedures. Industry consultants were used to determine the variables used to measure Australian dairy farm performance. The survey questionnaire was then developed, pre-tested and modified by three HRM faculty members, two industry experts and twenty selected dairy farmers, ensuring content validity [21, 22].

2.1 Data Collection

The Action Mailing Lists (AML)—the company that runs surveys—was employed to assist in randomly selecting 1549 dairy farmers out of a total of 6780 dairy farmers registered in Australia in 2010. The survey packs containing cover letter, consent form, and questionnaire were posted in August, with a follow-up in September 2010 [23]. This effort led to receiving a total of 205 useable responses (13 per cent response rate) for subsequent data analysis.

2.1.1 Explanatory Variables

The explanatory variables were specific HRM practices that were adopted by Australian dairy farm businesses in the sample. Sixty items of HRM practices were collected from the literature review and verified via industry focus group discussion and face-to-face interviews. These are believed to represent typical HRM features of small businesses in the farming industry [see 4, 8, 19, 24, 25]. All items were measured using a 7-point Likert scale, asking respondents to indicate the extent to which the item had been adopted in their farm (1 = strongly disagree;

7 = strongly agree). The Cronbach's alpha value of all items was .95, suggesting the internal consistency of the scale [26, 27].

Because of potential inter-correlation among 60 HRM variables, factor analysis was exploited to reduce a large amount of variables into non-correlated constructs suitable for subsequent regression analysis—the methods commonly used in the prior studies testing the HRM-performance link [2, 28]. Ten HRM practice factors were extracted from the principal component analysis with varimax rotation method. The value for Kaiser-Meyer-Olkin measure of sample adequacy is .820 (<1) and Bartlett's test of sphericity equals to 6,493.40 (df = 1770; sig = .000), suggesting the suitability of using the factor analysis. The Cronbach's alpha for all factors reached to .71, again indicating the internal consistency of the factor scale. The following is a brief explanation of each factor and how it was derived.

The first factor has a higher means on HRM practice items related to "selection of employees through formal job interviews" (0.821), "review of job applications" (0.723), "assessment of skills of potential employees" (0.800), and "use of reference checks" (0.725). All these items appear to relate to formal selection process, which was used by livestock managers in selecting their farm employees in recent years [see 29]. Hence the factor was labelled as '*formal selection process*'.

The second factor was loaded highly on the items related to "hiring using employment agreement" (0.811), "use of employment agencies for hiring employees" (0.621), and "use of standard hiring procedure for family members" (0.643). Therefore, the factor was termed as a "*formal hiring procedure*".

The third factor was related to the items such as "use of external training in the past 12 months" (0.809), "evaluation of training programs to measure the effectiveness to improve employees' competencies" (0.767), and moderately on "training for family workers" (0.570). It is widely discussed that, in the context of the farming sector, that the training process for farms mainly includes external training by industry experts, evaluation of training programs, and effectiveness of training on job performance [29, 30]. Therefore, this factor could be labelled as the "*training process*" as it combines the elements of external training opportunities, and evaluation of training effectiveness.

Factor 4 contains high loadings on "performance reviews of employees at least once a year" (0.855) and "performance reviews of family workers" (0.909). Prior studies concluded that reviews of employees' performance on an annual basis have been a well-established appraisal method in large firms as well as in small firms [29, 31]. Further, performance review of family workers in dairy farming could only be possible if the evaluation process is well established. Hence, Factor 4 was marked as "*annual performance review*" as it tied items relevant to performance reviews for both paid employees and family workers of Australian dairy farms.

Then, the fifth factor combines the related items such as "formal meeting with employees" (0.821), "involvement of family members in meetings" (0.793), "conduct of employee surveys related to farm issues at least once a year" (0.728), and "use of noticeboards for important messages" (0.629). Previous literature suggested having an open communication process with the involvement of employees in meetings and through the employees' surveys, so that employees can share their views and opinions [31, 32]. Therefore, Factor 5 is labelled as "*communication process*".

Also, it appears that "informal discussion on career aspirations with immediate boss" (0.722), "discussion on career planning of employees" (0.735), "other career opportunities within the dairy farm" (0.783), and "other career opportunities within the dairy industry" (0.856) are closely related. Hence, this combined Factor 6 was marked as *"career opportunities"*. It is argued that farmers were, in fact, quite proactive in suggesting career opportunities for their workers, especially within the dairy industry [30].

Factor 7 has high loadings on items related to "documented risk management process" (0.897) and "teaching employees to prevent farm accidents" (0.704). Guthrie et al. reported workers fatalities and injuries as a serious concern in the agricultural industry [20]. Hence, farmers could do everything possible to prevent OH&S issues such as farm accidents. The documented OH&S standards could help better manage the risks associated with injuries on dairy farms. Therefore, Factor 7 was labelled as *"risk management of OH&S issues"*.

The eighth factor loads high on items such as "farm regarded as a friendly workplace" (0.713), "social interaction among employees during work" (0.734), and "preference for daily informal social interactions rather than formal social gatherings" (0.787). Strochlic and Hamerschlag advocated the use of informal practices such as respectful treatment of employees by saying hello, inquiring about their families, sharing lunch with them, or inviting them into your homes for a meal [30]. The informal social interactions among the workforce, especially interactions between workers, supervisors, and owner-managers, are more likely to nurture relationships and strengthen social ties [8]. As a result, employee motivation may improve and reduce stress in the workplace. Factor 8 was, thus, named as *"informal social interaction"*.

Factor 9 emphasises employees' movements in their jobs with relatively high loadings on "workers compensation records" (0.796), "OH&S compliance" (0.677), "employees' leave" (0.831), "employee absenteeism" (0.811) and "employee turnover" (0.696). Previous researchers stressed the necessity for documentation of compliance-related issues in the farming workplace [33]. It is argued that a lack of compliance documentation would increase financial losses accompanied by possible legal action taken by employees against their employers. Bitsch and Olynk also stressed the importance of maintenance of formal HR records in the agricultural industry [29]. Therefore, this factor was related to *"compliance-related records"* as it emphasises common labour compliance issues such as compensation, employee leave records, absenteeism and turnover or termination.

The tenth factor has loaded high on items related to "milk harvesting" (0.887) and "maintenance and cleanliness of milk harvesting plant" (0.887). In the previous studies, the milking SOPs were considered important because of their effect on milk quality [4, 6]. These procedures are more likely to improve milking and cleanliness standards within a rotary milking shed. Thus, Factor 10 wan named as *"safe milking procedures"*.

The above-discussed ten HRM practice factors, such as formal and validated selection, hiring procedures, training process, annual performance review, communication process, career opportunities, occupational health and safety, informal social interaction, compliance-related records and safe milking procedures, were included in the subsequent regression analysis.

2.1.2 Dependent Variables

Initial farm performance variables were collected from the relevant literature review and advice provided by the Australian dairy farm experts and practitioners via the focus group and semi-structured interviews. Farm performance was measured by a number of variables, such as farm profitability, net farm profit shared with employees, morbidity rate, mortality rate, and percentage of cows treated for mastitis.

Farm profitability was measured, using two items: (1) perceived percentage increase in farm profitability in the past 12 months [34], coded as 2.5 to 62.5 by taking midpoints of their value ranges from 0–5 to $\geq$25 %; and (2) the extent to which net profit was perceived by respondents to have been shared with employees in the past 12 months, coded as 1 to 7 (1 = never shared to; 7 = always shared), respectively.

Herd health was measured, using three variables. First, perceived percentage of milking cows treated for mastitis in the past 12 months against the total number of cows. Their values ranged from 0–4 to >50 %. Second, perceived percentage of milking cows culled due to poor health conditions in the past 12 months against the total number of cows. Their values were also ranging from 0–4 to >50 %. Lastly, perceived percentage of cow deaths due to health-related causes in the past 12 months against the total number of cows. Their values ranged from 0–4 to >30 %.

Factor analysis was first used to subtract underlying factors of farm performance variables, but failed to meet criteria for factor analysis. It is concluded that factor analysis is not a suitable analytical technique to reduce farm performance variables into underlying factors in this case. However, it is argued that some of these performance variables measure similar concepts, hence they could be arbitrarily controlled as one factor. For example, first two variables "farm profitability" (0.573) and "net farm profit shared by employees" (0.779) were controlled and derived as one factor called "*financial outcome*".

The logic of using one-factor solution to derive a singular dependent factor of 'financial outcomes' can be explained as follows. As profitability is commonly used to indicate the financial performance of firms, it can be used to measure financial outcome (see previous studies [2, 5, 35]). It is likely that some dairy farms that have achieved high profitability would share net profit with employees. Therefore, higher perceived percentage of net profits shared with employees suggests a good financial outcome of the farm. Hence, these two variables were factored into one variable as "*Financial Outcome*" factor.

Similarly, the highly correlated variables such as morbidity rate (0.807), mortality rate (0.701) and the percentage of cows treated for mastitis (0.533) have commonly measured the broader construct, that is, the "*herd health*" in the context of dairy farming. It is argued that the morbidity rate, mortality rate, and percentage of cows treated for mastitis are conceptually similar [36, 37]. Hence, it is reasonable to apply a one-factor solution in this case too. Therefore, the factor is named "Herd Health".

In summary, two sets of several farm performance variables were analysed, based on similarity in concepts and one-factor solution. As a result, two factors, "*financial outcome*" and "*herd health*" were derived and used as dependent variables in the subsequent regression analysis.

The control variable "farm business strategy" (3 items) was measured by using a 7-point Likert scale ranging from '1' (strongly disagree) to '7' (strongly agree) with the statements in the survey instrument. Other control variables, such as "*herd size*" values ranging from 0–500 to >2,000 cows, and "farm age" ranging from 0–1 to >25 years were also included in the multiple regression analysis.

2.2 Data Management and Analytical Technique

Dairy farm businesses that did not provide either complete farm performance data or HRM data were removed from the final data set. Additionally, no outlier observations were found in the data set. The final data set for analyses contained 205 cases among Australian dairy farm businesses. The data were analysed using the SPSS 21. Means, standard deviations and correlation matrix of all relevant variables including HRM practice factors, farm performance indicators and contextual variables were first analysed, and presented in Table 1.

Ordinary Least Square (OLS) regression analysis technique was then adopted for the current study. OLS regression was used to test the effect of HRM practice factors on two farm performance indicators as indicated by financial outcome and herd health. The empirical models for regression analysis are presented in the following formula:

$$FINANCIAL\_OUTCOME = \beta_o + \beta_1 X_1 + \beta_2 X_2 + \beta_3 X_3 + \cdots\cdots\cdots + \beta_{10} X_{10} + \varepsilon$$

$$HERD\_HEALTH = \beta_o + \beta_1 X_1 + \beta_2 X_2 + \beta_3 X_3 + \cdots\cdots\cdots + \beta_{10} X_{10} + \varepsilon$$

where "X_1 to X_{10}" represent the HRM practice factors, "β" represents the regression coefficients, "β_o" represents the constant and "ε" represents the error.

3 Results

The analysis for the current research focused on examining the impact of HRM practices on dairy farm performance through OLS regression. To answer the research question, whether a set of HRM practices has any impact on farm performance as measured by financial outcome and herd health, two regression models (1–2) were run and presented in Table 2. These models assess the impact of HRM practice factors on financial outcome and herd health through their R^2 values and significance level. It is shown that each of the regression models has a reasonable

Table 1 Mean, standard deviation (SD) and correlation matrix of all measurements in the regression models

	Variables	N	M	SD	1	2	3	4	5	6	7	8	9	10	11	12	13	14	15	16	17
	HRM practice factors																				
1	Validated selection process	194	–	–	1																
2	Hiring procedures	194	–	–	−.01	1															
3	Training process	199	–	–	.24**	.43**	1														
4	Performance review	197	–	–	.05	.43**	.40**	1													
5	Communication process	201	–	–	.09	.45**	.40**	.48**	1												
6	Career opportunities	203	–	–	.28**	.32**	.38**	.37**	.43**	1											
7	OH&S risk management	203	–	–	.14	.26**	.33**	.27**	.34**	.30**	1										
8	Inf. social interactions	204	–	–	.24**	.07	.10	−.04	−.11	.20**	.07	1									
9	Compliance records	197	–	–	.35**	.06	.14	.03	.02	.15*	.11	.21**	1								
10	Safe milking procedures	201	–	–	.13	.17*	.08	.03	.09	−.06	.14*	.17*	.09	1							
	Farm performance variables																				
11	Financial outcome	194	–	–	−.09	.02	−.04	.23**	.22**	.06	.16*	.01	.01	.11	1						
12	Herd health	204	–	–	.17*	.01	−.02	−.12	−.08	−.02	.16*	−.13	−.05	.11	.09	1					
	Control variables																				
13	Product quality strategy	204	6.5	.7	.16*	−.01	.03	−.03	.02	.13	.16*	.26**	.19**	.08	.12	.13	1				

(continued)

Table 1 (continued)

	Variables	N	M	SD	1	2	3	4	5	6	7	8	9	10	11	12	13	14	15	16	17
14	Innovation tech strategy	205	5.4	1.4	.24**	.15*	.18*	.20**	.29**	.25**	.21**	.20**	.36**	.03	.01	.12	.36**	1			
15	People manage. strategy	205	5.9	1.1	.47**	.06	.22**	.08	.24**	.30**	.25**	.36**	.31**	.08	.07	.15*	.42**	.55**	1		
16	Herd size	205	1.2	.4	.17*	−.05	.06	−.07	−.14*	−.17*	.01	.08	.08	.07	.15*	.01	.06	.18*	.24**	1	
17	Farm age	205	5.3	1.1	.02	.19**	−.01	.05	−.03	−.02	.01	.18*	.09	.05	.02	.01	.21**	.20**	.02	.05	1

*** Correlation is significant at the 0.01 level (2-tailed)*

** Correlation is significant at the 0.05 level (2-tailed)*

Explanatory Notes*: Means and standard deviations (SD) of all factors are not suitable for tabulation*

Table 2 Multiple regression analysis—impact of HRM practices linked to business strategy on dairy farms' financial outcome and herd health

N = 205	**Financial Outcome**		**Herd Health**	
HRM practices factors	**Model 1a**	**Model 1b**	**Model 2a**	**Model 2b**
(Constant)[1]	−.96(.34)[1]	2.05(0.04)[1]	−.62(.54)[1]	−2.68(0.01)[1]
Formal & validated selection	−.17*	−.12***	.29**	.22*
Hiring procedures	−.29**	-.30**	.11†	.15†
Training process	−.24**	-.25**	–	–
Performance review	.29**	.29**	–	–
Communication process	.31**	.30**	–	–
Career development opportunities	–	–	−.07	−.16†
Risk management of OH&S	.19**	.22**	.21**	.17*
Informal social interaction	.11	.16***	.23**	.26**
Compliance-based HR records	–	–	–	–
Standard milking procedures	.16*	.21*	–	–
Control variables				
Business strategy—cost reduction		–		–
Business strategy—product quality		−.15***		.19**
Business strategy—tech innovation		–		.35**
Business strategy—people management		–		.33**
Compliance with workplace relations laws		.14***		.22*
Year of establishment—farm age)		–		–
Herd size		–		–
Model Fit				
R^2	**.23**	**.27**	**.18**	**.31**
Change in R^2	**–**	**.05**	**–**	**.17**
F value	**4.49****	**3.46****	**3.47****	**4.45****

[1] *"Constant" shows the t-values and their Sig. levels in parentheses. **p < 0.01, *p < 0.05, ***p < 0.1*

explanatory power as shown in R^2 values for *Financial Outcome Model* (1a) as equal to 0.23 and *Herd Health Model* (2a) at 0.18. These findings indicate that the adoption of a synergistic set of HRM practices significantly predict dairy farm performance in the areas of financial outcome ($p < 0.01$) and herd health ($p < 0.01$).

It is also found that most of the HRM practice factors, such as occupational health and safety, performance review, communication process, formal selection and hiring procedures, and informal social interactions have positive impact on farm performance indicators. This was revealed by examining the individual contributions of each HRM practice factor to farm performance indicators as indicated by their regression coefficients (β values) and significance level (see Table 2).

Specifically, the findings indicate that the "*annual performance review*" factor has a significant positive impact on financial outcome ($p < 0.01$). The regression coefficient shows that use of performance reviews in dairy farms increase financial outcome by 29 per cent.

The "*formal and validated selection*" ($p < 0.01$) and "*hiring practices*" ($p < 0.1$) were both found to have positive impacted on improving herd health, by 29 per cent and 11 per cent respectively.

Another HRM practice factor "*communication process*" significantly improved financial outcome ($p < 0.01$) by 31 per cent. This result seems reasonable because there is a general argument that effective communication may help farm workers to provide candid performance feedback, and make suggestions to improve production and to increase financial outcomes [38].

The "*risk management of occupational health and safety (OH&S)*" factor has moderate positive impact on both financial outcome ($p < 0.01$) and herd health ($p < 0.01$). Results suggest that risk management of OH&S is likely to be beneficial for dairy farm performance as it would increase financial outcome by 19 per cent and improve herd health about 21 per cent.

The "*informal social interaction*" factor was found to improve herd health ($P < 0.01$) by 23 per cent. But, how would such employee socialisation practices lead to improve herd health? This will be discussed further in next section.

In contrast, not all of the HRM practice factors contribute positively to enhancing farm performance. For instance, the "*training process*" was found to be negative and moderately related to financial outcome ($p < 0.01$). The "*career opportunities*" factor was negatively related to herd health, though the relationship was not strong ($p < 0.1$). The "*hiring procedures*" factor was also negatively related to financial outcome ($p < 0.01$). These interesting findings, though not overly surprising, deserve further discussion later.

In addition to the HRM practice factors, contextual variables were also included in the regression equations, and results are shown in Models 1b–2b (see Table 2). The contextual variables most significantly related to farm performance include those business strategies focusing on product quality, innovation technology, and people management. Other control variables such as years of establishment and farm size were found to have no or weak effects on farm performance. Nonetheless, the control variables account for additional variances in farm performance, beyond the combined influence of HRM practice factors. The change in R^2 value of 0.13 is particularly high for "*Herd Health Model (2b)*", indicating the significant influence of the contextual factors such as farm business strategy on improving herd health. Overall, HRM practice factors together with several contextual factors explain about 27 per cent of variance in *Financial Outcome* ($p < 0.01$), and 31 per cent in *Herd Health* ($p < 0.01$). We devote the next section to discuss the implications of findings generated from this study.

4 Discussion

The current study identified that several human resource management (HRM) practices were adopted by 205 Australian dairy farms investigated, and that these HRM practices were related to farm performance positively as well as negatively. HRM practices positively influencing farm performance cover performance review, communication process, formal selection and hiring procedures, informal social interactions and risk management of OHS. HRM practices negatively associated with farm performance include training and career development.

The most convincing argument of the positive impact of performance reviews is the setting of farm objectives and reviewing of employee's performance against those goals. The performance reviews tends to hold employees accountable for increased financial outcome. If an employee's performance remains below target after the review, other employees and owner-managers tend to put pressure on the poor performer to be more productive [33]. This is how performance reviews were conducted in the farming industry to motivate under-performing employees. Another argument is that a performance review may influence employees' competence, especially the upgrading of their technical skills. In addition, performance reviews could be seen as opportunities for employees to evaluate the gap between their current and expected skills, and to fill that gap [28]. This argument indicates that performance reviews are likely to enhance the technical skills of farm workers, for example, by enabling them to measure somatic cell counts, diagnose infectious diseases etc., and as a result, this practice would help improve herd health, in turn, increase financial performance.

Positive impact of open communication on farm performance is expected as open communication between employer and employees provides an environment to share critical information about farm tasks. Open communication helps employers to give feedback to improve employees' individual performance, which may contribute to the increased financial performance [39]. Prior studies also suggest that the information sharing through communication improves sales and firm profitability [39]. Furthermore, open communication tends to harness employees' sense of belonging and of being valued by farms, which could lead to greater employee motivation [30]. The motivated employees are more likely to perform their jobs in an efficient way in order to achieve farm goals. As a result, this would lead to better financial outcome.

The positive farm performance effects of risk management of OH&S are legitimate if farmers realise the negative outcomes caused by the lack of safety practices. It could be argued that the absence of OH&S practices would sometimes disrupt farm operations because of workplace accidents and injuries to key employees [20]. The accidents are also likely to trigger a vicious cycle that could drastically affect performance. Farmworkers suffering injuries from accidents at their workplace could demand extra compensation for medical expenses and health insurance [20]. In addition, farm accidents often occur because of ineffective OH&S practices, as a result, likely attract penalties from the government's compliance authorities [19]. Therefore, it appears that dairy farmers in the current study seriously considered their risk management of OH&S in order to achieve better farm performance.

The implementation of OH&S practices may also reduce the psychological hazards in dairy farming. The lack of OH&S practices can increase psychological hazards such as stress, fatigue, bullying, loneliness, and, sometimes conflicts arise from social interactions between the workers [29]. These hazards could be minimised by related OH&S practices including psychological counselling, paid holidays, utilizing family-friendly practices and creating opportunities for social interaction. Perhaps, informal social interaction could be the starting point to overcome psychological hazards, which would positively affect dairy farm performance.

This point was confirmed by the findings of the current research, which showed that informal social interaction improved herd health. Earlier studies suggest that social interaction by inviting employees for picnics, holiday dinners, pizza lunches, celebrations on special occasions, and counselling in personal matters play an important role in motivating employees in livestock farming [29]. It is projected that motivated employees are likely to become good carers for herd health. Furthermore, a positive social environment and informal culture helps facilitate the free flow of information and knowledge, create a sense of attachment, and build synergistic working relationships among employees, all of which would foster organisational learning and employee competency [28]. Subsequently, competent employees help improve production efficiency, while motivated employees tend to increase their personal interest towards their tasks. Farm hands with extensive social interaction might be the very type of workers who are able to diagnose the diseases more vigilantly; in turn, this may help improve herd health in dairy farming.

The results from the current study also found HRM practices negatively related to dairy farm performance. For instance, the negative effect of training on financial outcomes might be largely related to the nature of running small firms such as Australian dairy farms. Small firms often consider training as being of relatively limited relevance because of the high costs and time investment associated with this process [24, 40]. The argument is that training processes are likely to reduce financial outcomes as they are often treated as a cost item in the firm's balance sheet. Resource-limited and cost-conscious small firms such as Australian dairy farms tend to choose informal "*hands-on*" training practices, rather than lengthy training process.

Similar to the negative performance-effects of training, emphasis on employee career development opportunities may have negative effects on herd health. It could be argued that the farmworkers with alternative career opportunities were more likely to demand higher wages and extra on-farm benefits [33]. In such situations, if employees' demands were not considered, it might lead to low employee satisfaction and poor job performance, such as ignoring potential herd health issues. These issues, therefore, could be more likely to have negative effects on herd health.

We also found that formal hiring procedures, though positively related to herd health, but negatively related to financial outcome. There are two main reasons behind this intriguing finding. *First*, the farm managers may often not be willing to invest adequate time in the hiring process or they tended to undertake hiring without sufficient preparation [19]. *Second*, the farmers generally lack the skills required for an effective hiring process [33]. The inadequacies coupled with the time pressure to hire employees immediately forced farmers to forego and short-change the process [29]. These reasons, though justified by farmers, may induce a greater risk of hiring unsuitable employees without the required skillsets [19]. Subsequently, the new employees may have inadequate technical skills to improve farm performance. Employees without the proper skills would also likely quit their jobs leading to increased turnover costs, and lowering the financial performance [41]. These arguments may explain why the hiring procedures identified in the current study are negatively associated with financial outcome.

These findings do not imply that dairy farmers should not adopt recruitment and selection procedures. Interestingly, the formal and validated selection and hiring practices identified in the current study have positive impact on improving herd health, if implemented properly without time pressure. With sufficient preparation, and by a thorough review of job applications and selection criteria, dairy farmers could hire people with required skills. In addition, a new employee hired through such practices tend to take extra care of diseases like mastitis or lameness in the cows, which leads to lower rates of diseases, and the result is better herd health. Therefore, as illustrated in the current study, recruitment and selection practices, if implemented effectively, may have positive effects on herd health in dairy farming.

The most interesting results of the current study are those related to business strategies, taken as contextual variables, which were found to significantly affect dairy farm performance, both positively and negatively. For example, "*quality-focused business strategy*" can, in fact, lower financial outcome, though with a weak effect ($p < 0.1$). It is perhaps obvious that firms with a focus on the enhancement of product quality tend to invest heavily in the continuous training and development of their employees [42]. The increase in training, however, may help improve product quality, but it is likely to incur extra costs that, in turn, may affect the farm's financial return on investment.

In contrast, "*quality-focussed business strategy*" significantly improve herd health ($p < 0.01$) by 19 per cent. This is because farms with a quality-focussed strategy would have emphasised employees' training and development so as to have sufficient knowledge and skills for better herd care, leading to improved herd health. Schuler and Jackson argued that the quality-enhancement strategy with focus on training and development typically involved greater employee commitment and utilisation [14]. Sanz-Valee et al. also suggested that firms focusing on employee training would stimulate employee co-operation and obtain continuous improvement in quality [42]. Greater employee co-operation and commitment leads to positive role behaviours such as low absenteeism, in turn, may better take care of herd health in dairy farming.

Also, the "*innovation technology strategy*" in the dairy farming creates a positive influence on herd health. Innovative technology has been rapidly applied in the farming industry [43, 44]. Australian dairy farmers have also adopted various innovative technologies in the field of herd health, genetics, and feeding, which tend to improve their farm performance via increase milk production [45]. Furthermore, dairy farms following the innovation technology strategy are likely to achieve positive performance via selection of highly skilled employees, focussing on training, and performance management for long-term goals [14]. The constant adaptation to new technology is more likely to encourage employees to broaden their skillsets by taking risks and responsibilities, which may enhance employee work morale and commitment [14, 46]. In the context of dairy farming, such behavioural outcomes would have positive influence on improving herd health.

Finally, "*people management strategy*" was found to be most important in dairy farming. The concern of people management in business strategy might be due to the growing importance of paid employees along with family workers after recent structural changes in the dairy industry. Indeed, human resources should not be considered as a cost but as a critical source for achieving a firm's competitive advantage [47]. The argument determines that the effective management of people via a set of HRM practices works in favour of achieving farm performance [48]. So, it is clear from the findings of the current study that farm business strategies related to product quality, innovative technology adoption and people management, if implemented appropriately, were more likely to improve dairy farm performance.

5 Conclusions

This study expands the existing literature related to the impact of a synergistic set of HRM practices on dairy farm performance. Our purpose was to determine whether strategic HRM practices adopted by dairy farmers would affect their farm performance. The results empirically reaffirm an HRM-performance link. There exists an underlying relationship between HRM practice and performance, despite dairy farmers may choose different yet relevant set of HRM practices.

HRM practices, such as OH&S, performance reviews, communication process, informal social interactions, validated selection and hiring procedures was found to have a positive effect on farm performance, although, these HRM practices must be integrated to create positive effects and produce a greater source of competitive advantage [49, 50]. The empirical evidence also indicates that not all of the HRM practices (e.g. training and career development) are relevant in terms of improving farm performance, especially if dairy farms are operating in small scale. Strategically, it is reasonable for small businesses in the dairy farming to choose and adopt some HRM practices and to reject others in order to improve farm performance [51].

The findings of the current study have practical implications for dairy farmers. *First*, dairy farmers would now have a good understanding of the extent to which they have adopted HRM practices, and whether or which of these HRM practices are effective in improving both farm financial outcomes and herd health. For example, the dairy farmers are made aware of the importance of OH&S practices to reduce the risks of farm accidents and injuries, because such practices improve financial outcome by decreasing compensation costs related to medical and health insurance. *Second*, this study illustrated the importance of focusing on farm business strategies, when implementing relevant HRM practices. There were different effects of implementing different farm business strategies to achieve desirable farm performance with reference to financial outcomes and herd health. Dairy farmers need to choose appropriate business strategies and adopt specific HRM practices according to their particular environment and business settings.

References

1. MacDuffie, J. P. (1995). Human resource bundles and manufacturing performance: Organization logic and flexible production systems in the world auto industry. *Industrial and Labour Relations Review, 48*, 197–221.
2. Huselid, M. (1995). The impact of human resource management practices on turnover, productivity, and corporate financial performance. *Academy of Management Journal, 38*, 635–672.
3. Hoque, K. (1999). Human resource management and performance in the UK hotel industry. *British Journal of Industrial Relations, 37*, 419–443.
4. Stup, R., Hyde, J., & Holden, L. (2006). Relationships between selected human resource management practices and dairy farm performance. *Journal of Dairy Science, 89*, 1116–1120.
5. Hyde, J., Stup, R., & Holden, L. (2008). The effect of human resource management practices on farm profitability: an initial assessment. *Economics Bulletin, 17*, 1–10.
6. Hyde, J., Cornelisse, S. A., & Holden, L. A. (2011). Human resource management on dairy farms: Does investing in people matter? *Economics Bulletin, 31*, 208–217.
7. Huselid, M. A., Jackson, S. E., & Schuler, R. S. (1997). Technical and strategic human resource management effectiveness as determinants of firm performance. *Academy of Management Journal, 40*, 171–188.
8. Mugera, A. W., & Bitsch, V. (2005). Managing labor on dairy farms: A resource-based perspective with evidence from case studies. *International Food and Agribusiness Management Review, 8*, 80–98.
9. Ulrich, D. (2011). *HR Champions*. Boston: MA Harvard Business School Press.
10. TPiD (2012) The people in dairy program. Retrieved 17 February, 2012, from www.dairyaustralia.com.au/thepeopleindairy/
11. Kramar, R., Bartram, T., De Cieri, H., Noe, R., Hollenbeck, J., Gerhart, B., et al. (2014). *Human resource management: Strategy, people, performance* (5th ed.). New York: McGraw-Hill Education.
12. Nettle, R. (2012). Farmers growing farmers: The role of employment practices in reproducing dairy farming in Australia. Paper presented to the 10th European IFSA Symposium, Aarhus, Denmark, 1–4 July 2012.
13. Porter, M. E. (1980). *Competitive advantage: Creating and sustaining superior performance*. New York: Free Press.
14. Schuler, R. S., & Jackson, S. E. (1987). Linking competitive strategies with human resource management practices. *The Academy of Management Executive, 1*, 207–219.
15. Bewley, J., Palmer, R., & Jackson-Smith, D. B. (2001). An overview of experiences of Wisconsin dairy farmers who modernized their operations. *Journal of Dairy Science, 84*, 717–729.
16. Hadley, G., Harsh, S., & Wolf, C. (2002). Managerial and financial implications of major dairy farm expansions in Michigan and Wisconsin. *Journal of Dairy Science, 85*, 2053–2064.
17. Australia, Dairy. (2012). *Dairy 2012: Situation and outlook report*. Melbourne: Dairy Australia.
18. Searle, G. (2002). The reality of a career in the dairy industry: An employee's perspective. A survey of New Zealand Dairy Farm Staff, prepared for the Kellogg Rural Leadership Course.
19. Bitsch, V., Kassa, G. A., Harsh, S. B., & Mugera, A. W. (2006). Human resource management risks: Sources and control strategies based on dairy farmer focus groups. *Journal of Agricultural and Applied Economics, 38*, 123–136.
20. Guthrie, R., Goldacre, L., & Westaway, J. (2007). Workers compensation and occupational health and safety in the Australian agricultural industry. *The Agricultural Industry, 9*, 1–17.

21. Adams, J., Khan, H. T., Raeside, R., & White, D. (2007). *Research methods for Graduate Business and Social Science Students*. New York: Sage.
22. Creswell, J. W. (2009). *Research design: Qualitative, quantitative and mixed methods approaches*. New York: SAGE Publications, Incorporated.
23. Dillman, D. A. (2007). *Mail and internet surveys: The tailored design method*. New York: Wiley.
24. Kotey, B., & Slade, P. (2005). Formal human resource management practices in small growing firms. *Journal of Small Business Management, 43*, 16–40.
25. Wiesner, R., & Innes, P. (2010). Bleak house or bright prospect? HRM in Australian SMEs over 1998–2008. *Asia Pacific Journal of Human Resources, 48*, 151–184.
26. Nunnally, J. (1978). *Psychometric theory*. New York: McGraw-Hill.
27. DeVellis, R. (2003). Scale development: Theory and applications. In L. Bickman, & D. J. Rog (Eds.), *Applied social research methods series*. New York: Sage Publications, Inc.
28. Paul, A. K., & Anantharaman, R. N. (2003). Impact of people management practices on organizational performance: Analysis of a causal model. *International Journal of Human Resource Management, 14*, 1246–1266.
29. Bitsch, V., & Olynk, N. J. (2008). Risk-increasing and risk-reducing practices in human resource management: Focus group discussions with livestock managers. *Journal of Agricultural and Applied Economics, 40*, 185–194.
30. Strochlic, R., & Hamerschlag, K. (2005). Best labor management practices on twelve California farms: Toward a more sustainable food system. A report prepared by California Institute for Rural Studies.
31. Guest, D. E., Michie, J., Conway, N., & Sheehan, M. (2003). Human resource management and corporate performance in the UK. *British Journal of Industrial Relations, 41*, 291–314.
32. Way, S. A. (2002). High performance work systems and intermediate indicators of firm performance within the US small business sector. *Journal of Management, 28*, 765–785.
33. Bitsch, V., & Harsh, S. B. (2004). Labor risk attributes in the green industry: Business owners' and managers' perspectives. *Journal of Agricultural and Applied Economics, 36*, 731–745.
34. Faems, D., Sels, L., De Winne, S., & Maes, J. (2005). The effect of individual HR domains on financial performance: evidence from Belgian small businesses. *International Journal of Human Resource Management, 16*, 676–700.
35. Guthrie, J. P., Flood, P. C., Liu, W., & MacCurtain, S. (2009). High performance work systems in Ireland: human resource and organizational outcomes. *International Journal of Human Resource Management, 20*, 112–125.
36. Chenoweth, P., & Sanderson, M. (2005). *Beef practice: Cow-calf production medicine*. Oxford: Blackwell Publishing.
37. Halasa, T., Huijps, K., Østerås, O., & Hogeveen, H. (2007). Economic effects of bovine mastitis and mastitis management: A review. *Veterinary Quarterly, 29*, 18–31.
38. Marchand, L., Boekhorst, J., & McEwan, K. (2008). Human resource needs assessment for the Pork Industry. Report for Employment Ontario Project: Ontario Pork Industry Council.
39. Vlachos, I. (2008). The effect of human resource practices on organizational performance: Evidence from Greece. *International Journal of Human Resource Management, 19*, 74–97.
40. Gilbert, J., & Jones, G. (2000). Managing human resources in New Zealand small businesses. *Asia Pacific Journal of Human Resources, 38*, 55–68.
41. Zheng, C., & Wong, H. Y. (2007). Company training reduces employee turnover, or does it? *International Journal of Business and Management, 2*, 28–35.
42. Sanz-Valle, R., Sabater-Sanchez, R., & Aragon-Sanchez, A. (1999). Human resource management and business strategy links: An empirical study. *International Journal of Human Resource Management, 10*, 655–671.
43. Jago, J., Ohnstad, I., & Reinemann, D. J. (2007). Labor practices and technology adoption on New Zealand dairy farms. Paper presented at the Sixth International ASABE Dairy Housing Conference.
44. Dowie, C. (2012). Robotic rotary dairy far from ordinary, Farm Weekly. www.farmonline.com.au/news/nationalruralnews/

45. Lubulwa, M., & Shafron, W. (2007). *Australian Dairy Industry: Technology and farm management practices 2004–05*. ABARE Research Report 07.9 Canberra. Australia: Dairy Australia.
46. Bewley, J. (2010). Precision dairy farming: Advanced analysis solutions for future profitability. Paper presented at the First North American Conference on precision dairy management 2010, University of Kentucky, Kentucky, USA.
47. Barney, J. (1991). Firm resources and sustained competitive advantage. *Journal of Management, 17*, 99–120.
48. Teo, S. T., Le Clerc, M., & Galang, M. C. (2011). Human capital enhancing HRM systems and frontline employees in Australian manufacturing SMEs. *International Journal of Human Resource Management, 22*, 2522–2538.
49. Harel, G. H., & Tzafrir, S. S. (1999). The effect of human resource management practices on the perceptions of organizational and market performance of the firm. *Human Resource Management, 38*, 185–199.
50. Becker, B., & Huselid, M. (1998). High performance work systems and firm performance: A synthesis of research and managerial implications. *Research in Personnel and Human Resource Management, 1*, 53–101.
51. Subramony, M. (2006). Why organizations adopt some human resource management practices and reject others: An exploration of rationales. *Human Resource Management, 45*, 195–210.

Exploring Change in Small Firms' HRM Practices

R. Wapshott, O. Mallett and D. Spicer

Abstract The academic literature widely acknowledges changes and variation in the practices of small firms but only a small amount of empirical work has explored the processes through which HRM practices undergo change. Research has tended, instead, to examine the presence and effectiveness of HRM in small firms and has often viewed this in terms of a deficit model relating such practices to an understanding of HRM derived from larger firms. This chapter focuses on the recruitment and selection and staff payment practices in use in three small services firms to explore the everyday, ongoing detail of their HRM processes and practices. Identifying the different processes through which recruitment and selection and staff payment practices changed in the participant firms provides a base for discussing persistent forms of informality and the lack of stability that reflects the everyday realities of the firms, not only in contrast to their formalized policies but in engagement with them. This chapter advances understanding of selected HRM practices in small services firms after periods of formalization and adoption of HRM policies and practices. The chapter also discusses how developing knowledge of small firms' HRM practices in this way has implications for researchers and practitioners.

1 Introduction

Small firms are frequently more informal and ad hoc than their larger counterparts, subject to turbulent internal and external environments in which they encounter complex, interacting forces influencing their operation such as employees, clients

R. Wapshott (✉)
Management School, Sheffield University, Conduit Road, Sheffield S10 1FL, UK
e-mail: r.wapshott@sheffield.ac.uk

O. Mallett
Durham University Business School, Durham University, Millhill Lane, Durham DH1 3LB, UK

D. Spicer
School of Management, Bradford University, Emm Lane, Bradford BD9 4JL, UK

C. Machado and J. P. Davim (eds.), *Work Organization and Human Resource Management*, Management and Industrial Engineering,
DOI: 10.1007/978-3-319-06376-8_5,

and markets [1–3]. As small firms grow it has been argued that they introduce greater degrees of formality, for example in terms of more sophisticated Human Resource Management (HRM) policies and practices [4].

However, there are relatively few detailed studies considering the processes of how this HRM is adopted and may itself be subject to change as firms respond to turbulent operating contexts. This represents an important gap in research concerning the nature of HRM activities in small firms because static snapshots of these policies and practices may misrepresent, for example, the degree of ongoing change and lack of stability in their adoption, implementation and maintenance. Drawing on an 18 month case study of three small firms this chapter focuses on the firms' recruitment and selection and staff payment practices to explore these issues in detail. In doing so, we seek to develop understanding of small firms HRM practices and offer a broad characterization of changes in these practices in relation to persistent forms of informality relating to environmental turbulence, employment relationships and owner manager prerogative.

2 Literature Review

2.1 HRM in SMEs

One of the central themes of research into small firms' people management activities is acknowledging the heterogeneity of this group of organizations. The diversity of firms that may be gathered under the 'SME' banner underpins frequently identified definitional difficulties [5] but also raises problems when seeking to compare firms' practices. A medium sized firm of 240 people is likely to differ significantly from a small firm with 24 employees. The influence of owner managers, business objectives, stage of organizational development and the nature of work undertaken in organizations have all been discussed as ways in which organizations may differ in how they approach HRM. In particular, the often related considerations of owner manager influence and organizational objectives demonstrate how simple assumptions around management control and profit maximization may not apply to all, or even a majority, of small firms. When discussing SMEs or, as in our focus, small firms with fewer than 50 employees, it might not be possible to identify every way in which firms differ but it remains important to understand which differences may be more or less important and therefore whether certain firms may in fact be comparable.

Furthermore, there is a risk in discussing 'HRM' in the broad SME literature that it is considered from a deficit position; it identifies what small firms are not doing or have not got in place when compared against a formal set of 'HRM' practices and policies. This is at odds with more general conceptualizations of small firms which recognize that comparisons between small firms and their larger counterparts do not represent the best way to understand these firms [6, 7]. Adapting conceptualizations of HRM processes from the large firm literature is an implicit way of holding the same comparison. Where existing studies often focus on debating the presence of HRM in smaller firms (for example, [7]), we suggest that the debate should be less about whether or not some form of HRM exists in small firms but more about what functions it performs and how or when it is utilized. For example, such studies of HRM practices could focus on why they are there, how they work, how they reflect or protect particular interests and at what cost or benefits.

In attempting to understand the dynamics of HRM practices it is important to acknowledge that small firms may not only differ from one another but also differ from larger firms in several important ways [8, 9] that may significantly alter how such practices are adopted and put into use over time. For example, the development, or emergence, of HRM practices and routines is subject to varying degrees of influence from external environmental factors. Rainnie's [10] typology of relationships between small firms and their larger counterparts emphasizes the vulnerability of some small firms and the implications this could have for their employment practices. The potential vulnerability of small firms to competitors and in particular to their clients' demands might be explained, in some instances, by their relative 'resource poverty' [6, 11] which leaves them unable to absorb shocks from the market. However the nature of client influence in any given instance is known to vary [12–14]. So, while clients are acknowledged as playing a potentially important role in shaping small firms' HRM practices, the varying nature and extent of any influence makes their impact hard to generalize.

Small firms, further, are potentially less likely to have collective bargaining or trade unions [15], they lack internal labor markets [16] and tend to have more centralized decision making [17]. Management practices in small firms are acknowledged to be less formal and less strategic than those found in larger businesses [18]. Their cultures are shaped, at least initially, by owners whose goals and desires are communicated directly to employees in a context of close spatial and social proximity [19, 20], which also fosters informal relationships and working practices that are generally flexible and quick to change and adapt [21, 22].

The owner managers of small firms are traditionally acknowledged to exert influence over the practices in use [20, 23]. However such influence can be tempered by the assertion of employee interests; employees are not passive recipients of owner manager control. The presence of family or social ties in a workplace [24] and the relative importance of an employee's role to the business operation [25] can shape the employment relationship. As a result, HRM within small firms can become a product of informal negotiations and mutual adjustment between owners and workers [26, 27], creating more complex and subtle relationships than are commonly recognized in assumptions of owner manager dominance [28, 29].

Whilst such differences are a matter of degree and not kind (see page 238 in [30]), this relative informality [3, 31] can be clearly seen in respect of HRM activities [18, 32], for example relying on an owner manager's personal networks for staff recruitment, informal evaluation of training needs and pay setting processes [33]. Verreynne et al. (page 406 in [34]) argue that the "scarcity of resources drives diversified and divergent employment systems" in small firms, especially in their early stages of development. Within such a context, HRM practices are considered to evolve such that ad hoc practices become informal routines for addressing various organizational challenges [35] and there is a general expectation that formality in HRM practices will increase as businesses grow [32, 36], enhancing their competitive advantage [37]. Such formalization has been widely related to business growth and a sense of maturation [38].

HRM practices in small firms can be seen as products of not just single, isolated factors but also interactions between multiple influences. There is a need to understand HRM practices in small firms as they develop over time, understanding these everyday, ongoing processes in their own right rather than how they are deficient in relation to larger firms. The impact of these interactions and ongoing processes can be explored in relation to two aspects of HRM practice that are necessary to any commercial firm: recruitment and selection and staff payment.

2.2 Recruitment and Selection and Staff Payment

Recruitment and selection is an area identified as problematic for many small firms because, while an owner manager may reserve the right to select who they wish, they might not be able to attract the very best workers in a given field [39] or find sufficient employees with the requisite level of skills [40, 41]. The reasons for such problems vary but may include the levels of pay on offer [24], the state of the labor market and the lower levels of perceived legitimacy for smaller firms as employers [42].

In response to such difficulties small firms may reject such formal 'sophistication' [7, 43] and focus on tried and tested methods to maximize information on applicants before making a selection decision [11, 19]. Techniques include taking advantage of competitors or other firms making redundancies [44, 45], exploiting networks [46], staff recommendations [40], employment agencies [40, 44], advertising [47] and word of mouth [11]. The nature of recruitment and selection practices might therefore be described as often informal and reflecting the wishes of an owner manager but moderated by the position occupied by some smaller firms within the wider labor market. Whilst there is an assumption in the literature that such practices become more formal as firms grow [38], not enough is known about the ways in which such formalized practices change in response to other pressures or during shorter time frames.

Pay arrangements in small firms may also be characterised by their informality and the degree of discretion accorded to the employer within the context of external influences [48, 49]. Within this employer dominated context there may be scope for employee consultation but it is not always apparent how far these employee views influence salary levels [27, 50, 51]. The primary importance of owner managers in setting basic pay levels is also apparent in relation to contingent rewards, where bonus payments represent another area of management discretion [48]. More generally, however, contingent forms of employee reward are relatively uncommon in smaller firms [52, 53] not least because of difficulties associated with their adoption [11]. The degree of owner manager discretion associated with pay practices in small firms suggests that they can be changed if the owner wishes.

Both recruitment and selection and staff payment can be seen as potentially subject to change, albeit driven by different blends of influences in the firms' environments. The relative informality of practices in use and complex interactions of influences shaping HRM practices in small firms can result in an appearance of adhocracy [27] regarding employment matters, although the ad hoc nature of these practices may reduce through routinization [35] or through formalization as firms grow [36, 38]. What the existing literature does not currently explore to any great extent are the processes of how changes in formalized HRM practices occur. In other words, existing knowledge highlights that change occurs and indicates the factors that may serve to bring this about [53, 54] but there is little insight on how these changes operate, how the practices in small firms are formed and reformed over time [49]. This chapter presents and discusses findings on the processes and experiences of changes in recruitment and selection and staff payment practices at three relatively formal small firms.

2.3 Small Professional Service Firms

Professional service firms are interesting for HRM research because highly skilled, specialist employees are integral to service transactions [55] and part of the interface with clients [56]. They are therefore relied upon to generate income [57] by creating bespoke, intangible solutions to complex client problems [page 610 in [58]. How such relatively autonomous employees [59] are managed therefore impacts greatly upon business outcomes [60].

Unlike their counterparts in less skilled small firms, specialist workers in small professional service firms are likely to have a potentially stronger bargaining position in shaping their employment relationships [61]. Their employers may therefore be required to manage these valuable resources in different ways from lower skilled and more abundant workers [62–64]. HRM practices can be used to moderate the relationship between these organizations and their workforces [65, 66] and more formal practices have been identified in specialist small firms than the somewhat informal and 'unsophisticated' picture presented elsewhere [43, 67].

They therefore represent a useful type of small organization to explore the enactment and utilization of formal HRM practices over time. Research exploring the relative informality of smaller firms, together with the particular type of influence held by specialist workers, has the potential to produce different insights into the adoption and development of HRM in practice. However, the development of working practices and HRM in small professional service firms and other small, knowledge intensive organizations remains relatively under researched, despite their importance for many economies [30].

3 Method

This research set out to explore in detail the employment relationships and HRM activities of three small professional service firms. We adopted a multiple case study strategy [68] with the data collection concentrated into two main phases, approximately 18 months apart and covering up to three weeks at each organization. Contact was maintained with the firms between the main data collection phases. In addition to meetings with management, email and telephone interactions with employees helped to maintain relationships and deepen knowledge of practices and general workings of the companies. Incorporating a longitudinal component [48, 53] into the research design was important to identifying any changes in practice occurring.

3.1 The Firms

For this study businesses were selected with fewer than 49 employees and falling within Morris and Empson's [page 610 in 58] broad definition of a professional service firm as one "that trades mainly on the knowledge of its human capital, that is its employees and the producer-owners, to develop and deliver intangible solutions to client problems…the products of the PSFs are generally more complex [in comparison with other service companies] and, crucially, can be bundled in a variety of ways to customize solutions to the client's specific circumstances." Two of the participating businesses were specialist recruitment companies and the third was a design and communications company in which one of the founding directors was Chartered Institute of Personnel and Development (the UK's professional body for HR practitioners) trained and qualified giving these firms greater formal HRM knowledge than is often associated with small firms [69].

The firms were accessed through personal networking among business contacts [25, 28] and targeted mailings. These approaches proved effective in overcoming the problems often associated with accessing small firms for research purposes [70]. These three firms were identified as having relatively formal HRM policies and practices, from some perspectives to have 'formalized' [4]. Initially the recruitment businesses (FinRec and SciRec), which might be viewed as having business processes amenable to standardization, were complemented by two design and communications businesses, which might be anticipated to be more focused on innovation and non-standardized approaches to their work. However during the research period one of the design and communications firms out grew the business selection criteria and so is not presented here.

The businesses are identified by the pseudonyms FinRec, SciRec and ComCo and can be described as follows.

3.1.1 FinRec

FinRec Limited specializes in the recruitment of permanent staff to leading players in the financial services and consumer credit industries. The precise nature of the service provided by FinRec depends on client requirements but can involve recruitment of individuals for replacement purposes or managing a project to recruit an entire team. When this research commenced, the business had been established for around 10 years. It was founded by the managing director, Graham, who had worked in retail finance prior to establishing FinRec.

During the research period the business employed between 12 and 16 staff, the majority of whom were consultants providing recruitment services to clients. Other staff roles included Sally, the operations manager, and some administrative support. Employees came from a variety of backgrounds including a former HR manager and university graduates as well as people with relevant work experience but without higher education qualifications.

3.1.2 SciRec

SciRec Limited is a contracted supplier of specialist recruitment services to scientific businesses. Additional, related services are offered such as psychometric testing and interview training. Tim, the owner manager, had spent his career working in the pharmaceutical industry fulfilling a similar role to that served by SciRec. The business had been operating for four years when this study started.

At the time of the initial fieldwork SciRec employed an operations manager, two recruitment consultants focusing on technical and non-technical roles respectively, and two administrators. One of the administrators was on a fixed term contract and was the only member of staff without some recruiting responsibilities. During the second fieldwork phase five people were working for the company, however these were different from those employed in phase one. There was a recruitment manager (Chris) who fulfilled a very similar role to the one previously performed by the operations manager, a generalist recruiter who covered all non-scientific vacancies (Carly) and two scientific recruiters. One of these recruiters had a specific business development remit. One scientific consultant (Bethan) and the business development consultant were on initial fixed term contracts so that their abilities could be assessed before permanent appointments were made. The organization also employed an administrator. Consultants and management staff were mostly graduates.

3.1.3 ComCo

Finally, ComCo is a design and communications consultancy offering public relations, strategic marketing, design and internal communications services. ComCo was in its third year of trading full time when the research started, although the company had been established for five years as the directors prepared to leave their respective employers. Three founding directors each owned shares in the business with two of them, Fred and John, each holding 35 % of the business. The remaining 30 % stake was evenly divided between the third founder and a long-time associate (Alan) who joined ComCo as a director.

Although originally serving local authorities and small businesses, ComCo focused increasingly on multinational companies with some success. The changing nature of its business had implications for ComCo's structure and there was some staff turnover including John who left following disputes over the business' direction. Along with some staff turnover, headcount at ComCo increased from 14 to 20 during the study divided across the core of PR/communications team and smaller teams for marketing and graphic design. The growth in staff numbers was accounted for by growth in the marketing and design functions.

3.1.4 Data Collection

Data gathering in the three companies included direct observations (348 h), semi structured interviews (35) and reviewing company documentation (600 pages). The observations were used to gather data around employment relationships and the practices of managing employment relationships in each firm, beyond the formal and partial picture presented in written company HRM policies [52]. Drawing on such observations not only facilitated understanding of the research environment but also stimulated further questions and provided common ground to refer to in the semi structured research interviews with staff. One example of this was using observations on an incentive game at FinRec as a way into discussing performance management and targets with company employees. Additionally, this approach gave some scope to explore the potential for policy/practice tensions and contradictions that can be overlooked, for example in survey research.

Semi structured interviews were used to balance consistency across interviews while allowing some scope for discovering any informal or idiosyncratic approaches to HRM issues. The interviews, typically lasting 60 min, covered topics including how staff were hired and payment practices in the firm with a view to understanding the employment practices in each company. Care was taken to interview people who might be able to provide insight on specific topics, for example speaking to new starters who could talk about their experiences of recruitment and selection and staff induction.

Interviews were recorded with the permission of the interviewee and transcribed for analysis. Originally it was intended that interviewees from phase one would be re interviewed in phase two of the data collection however staff turnover, absences and the available research access limited the possibilities for such follow up interviews. Nevertheless, there were still sufficient opportunities to discover which practices were in use at different times of the study. Further, the high degree of change within the firms became a core area of focus within the study.

Informal discussions provided another means of data collection and were particularly important in FinRec during phase two because the owner manager was unwilling to release staff for further formal interviews. In addition to the single semi structured interview that was possible, staff were happy to discuss the company and their roles over lunch breaks and during quiet periods of work time. Notes from such interactions were written up immediately after the conversations to capture the details and then incorporated into the dataset for each business.

The observations and interviews were further bolstered by company documents. Initial discussions with the case study firms confirmed that, consistent with other SME research [15, 71, 72] and the training of those within the companies, each had some formal people management policies so documents including HRM policies, internal memos and business plans were gathered where available.

3.1.5 Data Analysis

Data analysis was guided by Miles and Huberman's [73, 74] procedures for qualitative data analysis because these sourcebooks set out approaches to enable thorough and systematic analysis of qualitative data. Throughout the data collection phases the observation, interview and documentary data gathered were read and considered against existing data from this study and extant literature to stimulate thought and highlight shortfalls in the data collected up to that point [74]. Further analysis was conducted at the end of the data collection phase via systematically applying pre-defined codes to interview transcripts, observation records and company documents to identify HRM practices in use.

Developing pre-defined codes, informed by the fieldwork, was driven by a need to order the data. The codes were designed to identify HRM practices around recruitment and selection, training and development, performance management, salary/pay, appraisal and staff exit, broadly following descriptions of HRM found in the work of Galbraith and Nathanson [75] and Devanna et al. [76] among others. Practices conforming to rigid definitions of HRM practices may not be found in smaller firms [7] so in this study the codes were interpreted broadly but consistently to permit comparison of practices in use. Coding was conducted by this chapter's first author while the codes and coding were checked independently for meaning and consistency by the chapter's coauthors.

3.2 Findings

While the literature suggests that some small firms' turbulent environments and informal practices will give rise to changes, it is the processes of how changes occurred that are of specific interest to this chapter. While our research explored a broad spectrum of HRM practices and areas of the employment relationship in the firms, two areas of HRM activity (*recruitment and selection* and *staff payment*) have been selected to provide sufficient scope for exploring the practices in use, rather than cover a wider range of practices in less depth. These findings therefore present examples of the ongoing, everyday changes within the formal HRM practices in the three firms.

One of the first things noticeable about the HRM practices found within these firms is that they are relatively sophisticated compared to what might be anticipated from many small firms [11, 45, 51]. For instance in recruitment and selection, SciRec and FinRec report the use of some psychometric testing for staff selection or multiple stages of selection. While ComCo also adopted systematic and structured approaches to staff recruitment and selection and utilized head hunters to target suitable talent for the business. The firms, as providing professional services, might perceive a need to deploy relatively sophisticated HRM practices given the importance of staff to their businesses' success [61]. Clearly, the complexity of recruitment and selection practices in the companies can be understood in terms of their business specialism (FinRec and SciRec) or training of an owner (John at ComCo). The owner managers' backgrounds, coming from large organizations with management practices that they imported into their own businesses, might also explain something of the apparent sophistication of practices in the firms. It might also lend weight to our impression that the formal policies were implemented sincerely to shape business practice rather than simply as 'window dressing' for clients' benefit [77]. Another area of similarity between the firms, but one in which they were also more typical of small firms generally, was in the changing nature of their practices.

Where staff payment practices were concerned, in all three firms pay rates were determined to varying degrees at the discretion of the owner manager or management team, consistent with what might be anticipated from extant literature. Nevertheless, FinRec and SciRec offer evidence of performance related pay schemes, which may be considered unusual in the context of small firms [11, 21]. The approach of ComCo to staff payment was interesting in its own way so far as it appeared closely tied to the company's development towards working with blue chip clients.

During the 18 months covered by this study, each of the firms displayed changes in their formal recruitment and selection and staff payment practices. We present and discuss these findings in terms of the three key themes that emerged in our analysis of the change processes relating to HRM practices: enduring, adaptive change in response to external operating environments, fluctuating, continuous change and short term periods of trial and error. In presenting these three themes our aim is not to categorize types of change in small firms' HRM practices but to seek to characterize the different processes through which such changes come about.

3.2.1 Enduring Adaptation

Enduring changes, in so far as can be identified from this study, emerged in ComCo's recruitment and selection and payment practices through adaptive responses to external pressures and became established features of its approach to

HRM. These changes were initiated during the first phase of onsite research which founding owner manager John described with an emphasis on cost control regarding staff salaries:

> In taking people on it's always about, you know, the balance of risk and reward. You know, you're committed to pay them a certain amount of money which puts pressure on the business to earn more money.

Such concerns over employment costs were balanced by a need for salary levels to retain good staff in a competitive client and labor market:

> When we get them, people, to the level we'd like, we are paying them above [the average]. This is a business [industry sector] where people move around a lot. We want to keep people because they, because it's such a relationship based business, if you were to lose them you are in danger of losing the client because the client likes to see that person. (John)

However 18 months later the business had undergone significant changes, reflected in ComCo's approach to staff payment. Tensions arising in the business had reached a breaking point. John had left the business in a disagreement over the direction it was taking and the demands this placed on his work life balance, and his original business partner Fred had become the managing director. Although company director Alan coyly described this process as being "as amicable as it could be", others, including Fred himself, acknowledged that John's exit had been a strained process. Following John's departure the original caution regarding wage costs had disappeared in the moves to employ staff with greater experience who were better able to meet the needs of ComCo's new clients and who could work independently of the management team. Fred explained that the change in focus had increased the wage bill by around 25 % but he asserted that the relatively high fees earned from large clients justified the increased costs. Thus, the approach to pay was largely dictated by the bargaining power of employees trusted by external clients who were in turn supported by what were perceived by Fred as the pressures of the external labor market.

At FinRec, consultant grade staff received a basic salary plus commission on revenue earned above an accepted threshold. The threshold varied with staff grade so senior employees with higher basic salaries had to generate more money than junior staff before they started to earn commission. FinRec operations were designed to deliver certain levels of financial returns to managing director Graham and, in addition to influencing recruitment and selection activities, this also shaped contingent pay arrangements. During one of several conversations with the researchers between the main onsite research phases, Graham and Operations Manager Sally explained how two of FinRec's major clients froze their recruitment budgets when experiencing a downturn in trade. In response, FinRec rapidly made three redundancies and raised the commission threshold to keep the business on track for Graham's financial targets. For the majority of consultants this meant having to earn the business an additional 12 % of revenue before they received any commission.

Although this change in pay arrangements appeared significant, staff were already accustomed to their targets being raised. In relation to a further target change, the staff seemed resigned to accept Graham's authority to alter commission without consultation: "at the end of the day, he's the boss that means what he says goes." (Mary, Consultant) and "You had your moan and that was it. Well, it's gotta be done so, I think everybody had their [grumbles] but there's nothing we could do about it" (Lorraine, Consultant). Even as the clients started spending again, however, the changes remained in place. For the staff at FinRec, the changes in pay arrangements were seen as part of normal business and Graham's prerogative.

In some respects adaptations such as these might be viewed as the successful (if somewhat opportunistic) improvizations or trials, those sustainable responses to shifts in firms' operating environments. Certainly both organizations changed in response to client related issues but the changes can also be seen as consistent with the respective organizational strategies and power within each business consolidated in the owner manager's prerogative. ComCo wanted to attract and employ staff able to serve the new client base it was winning, and FinRec needed to keep delivering returns for Graham. The changes and, importantly, maintaining them can be seen as high priorities of those running the firms. Moreover, perhaps reflecting this priority status, there was management resolve to see these changes succeed. These appear close to a fairly strategic approach to HRM activities but were not representative of all practices in these firms.

3.2.2 Fluctuating Practices

Evidence from SciRec suggested that, despite possessing sophisticated procedures for recruitment and selection and using them at various times, practice in this area shifted without discernible rationale. Although owner manager Tim explained the almost continual change as resulting from business pressures, staff hired at different times when such pressures were absent still reported various approaches to their recruitment and selection experiences.

Despite possessing the most seemingly sophisticated selection tools of all the companies studied, Tim described how formal employee selection processes had been relaxed during a period of high client demand. One specific consequence, attributed by Tim to this relaxation, was a consultant being appointed who was unsuitable for the firm and, as a result, Tim resolved to use the battery of tests and techniques available:

> So after that we said "OK, we're going to put them through a process" so we now put everyone through at least [psychometric test] plus at least two interviews, possibly a presentation, they get to talk with the team, you know to have a 10, 15 minute chat with each of the team, to try and make sure we get the right people 'cause we're saying to our clients "You need to put people through a more robust process" and we weren't, so we need to sort of practice what we preach

However employee accounts gathered in phase two revealed that practices continued to alter, even outside busy periods cited by Tim as causing the earlier shifts in practice. The consultants described their different selection experiences. Carly described her experience:

> I came in and spoke to Chris [recruitment manager] for half an hour, then spoke to Tim for I suppose about half an hour. The feedback I got from that was that they weren't sure I was confident enough for the position...So I was invited in for a second interview [including presentation to the team] to see if I could prove them wrong and, I don't know, for lack of other candidates or whatever, I got the job!

Others explained how, in addition to interviews, they had been placed on a temporary contract by way of indefinite trial period. Bethan (consultant):

> The first interview was really, Chris gave me an introduction about what SciRec does, and what they were looking for. He emphasized the recruitment side and also the sales side as well, erm, and then he was generally just sort of asked me questions with regards to my personality and I just sort of told him how brilliant I was! The second wasn't really an interview it was just sort of signing a contract, which actually turned out to be a temp contract, not a permanent contract...

The fluctuating practices detected at SciRec may indicate that Tim's sophisticated practices were unsuitable for such a small business [47] and may indicate some of the problems associated with trying to balance informal and formal processes in a small firm [48]. With Tim managing the business on a daily basis it is understandable that he may lack the time to implement the formal processes that he used in support of his successful Investors in People accreditation and recommends to much larger client organizations. As a result, the informal, ad hoc approach associated with small firms before they formalize and adopt HRM policies and practices persisted in SciRec, despite the sophisticated tools at their disposal and adopted when circumstances allowed.

The pressure to 'practice what we preach' without the resources to do so may help explain Tim's dogged insistence on formal processes even if they are subject to persistent informality and continual change. Another example of informally driven continuing change was found where Tim revised the commission structure several times within the previous 12 months. Tim was attempting something many small firms struggle with, designing an effective performance related pay system [11]. A desire to establish a complex reward structure while seeming to lack the necessary knowhow may underlie the ongoing adjustments to consultants' commission arrangements. Tim's actions created an impression of ad hoc or almost improvized approaches to these areas of practice. Closer consideration identified a frustration at the disparity between his desire for consistent, sophisticated formal practices and his ability to realize them in everyday practice.

Such continual change, without apparent pattern, may be characterized as fluctuating and this form of change was often attributed to external pressures, demands or opportunities. For example, a firm may adopt a practice dictated by a large client which does not become an embedded practice but, rather, is soon discarded or ignored. Owner manager centrality in the firms was such that, despite high degrees of formality, informal, ad hoc decision making persisted in response to pressures, irrespective of written HRM policies. They also suggest that, with such continual changes, the prospects for gradual routinization of practices are low.

3.2.3 Short Term Trial and Error

At FinRec, despite some enduring, adaptive changes, influential forms of informality also persisted in altering the ongoing, everyday HRM practices. Staff requirements were driven by managing director Graham's budget for the year. Graham calculated the required headcount by dividing his financial target by his view of what revenue each consultant should generate, as operations manager Sally explained:

> We've got at the end of the year, erm, a figure that Graham wants to achieve. To ensure that we get to that figure we need X amount of people within this business producing. So that's kind, that's the kind of goal as well. You've also got to see and plan for people who may leave. It's not a job whereby you start one week and the next week you're earning [revenue for the company]

FinRec staff were usually selected for employment at the end of a carefully organized process, features of which were copied from their clients' business practices. As explained by team leader Maggie, often the first contact made by FinRec with an applicant was by phone. A brief conversation assessed whether they might be suitable for interview, the principal selection tool. Two interviews were held, first with the two team managers and then a second round with either Sally (the operations manager) or Jenny (the managing director's wife and ad hoc employee).

The selection interviews normally took place on different days and included a "branch visit" when the applicant would shadow a consultant for a short period by way of job preview. Consultants described how some of them had sat selection tests, for example an in tray exercise or a job related personality test. The selection process did not end on appointment however, as staff unable to achieve their targets within the six month probationary period were usually dismissed.

However, employee selection at FinRec did not always occur in this way. Sally explained that when a number of consultants had left the business in quick succession FinRec had insufficient headcount to remain on target for revenue generation. Graham responded by exerting pressure on Sally to fill these positions rapidly, which had implications for the selection processes. Sally compressed both personal interviews and the branch visit into one day and lowered the standard of applicant appointed. Sally explained how she had acted in order to meet rapidly the headcount requirement: "Sometimes you do that, you go into panic mode."

Having overseen the under performance and dismissal of staff appointed hastily, Sally reflected that FinRec would have to return to a more controlled selection approach:

> I'm conscious now, we need another three people. I could quite easily have thought last week: "They're borderline these two, let's just get them in, let's get them in". But I'm not doing that this time. It might affect revenue but then I think: "Well what's the point of getting somebody in for a couple of months and then I have to get rid of them?" because it's not fair to them. It's not fair to us.

Changes brought about by informal pressure from managing director Graham to the recruitment and selection practices were introduced reflecting a degree of owner manager prerogative but then quickly reversed when the new system led to what were perceived as poor selection decisions. Departures from established practices or routines were short lived as undesired consequences caused a reversion to those established practices even though they were also recognized as having some associated costs. Such short lived changes, reverting to established practices, may be characterized in relation to persistent informality engaging over time with formalized, written policies and associated practices.

At SciRec, Tim's most recent changes to commission structures came after a previously unsuccessful attempt at revising them. On that occasion the consultants in post resisted his new approach for fear of losing money. During the first research phase there was an intricate combination of contingent pay arrangements in operation at SciRec. Tim described these schemes as a reward and an incentive for filling client vacancies, which earned revenue for the firm. However the consultants perceived inequity in the system because, under its complex rules, certain vacancies attracted commission while other, similar ones did not. Tim's initial attempt to impose changes to the performance related pay schemes was successfully opposed by the consultants who argued they would lose hundreds of pounds each month under the revised system. As a result, on this occasion, pay arrangements reverted to the previous formal system.

These practices at FinRec and SciRec may highlight the use of 'trial and error' in adjusting formal practices in small firms. In both instances changes were aborted once greater problems emerged than those they were seeking to overcome. Given the speed at which the consequences of an often informal, adhocratic decision to alter recruitment and selection or staff payment practices occurred, the nature of changes to practice represent a lack of stability and strategic direction even though a new system may eventually be arrived at. While such enduring changes may, eventually, result from a period of routinization this process appears turbulent and responsive to informal influences and forms of negotiation and mutual adjustment [24, 27].

For FinRec and SciRec the owner managers often asserted that there was a clear understanding of what their businesses needed in order to succeed but that there were barriers to succeeding in the changes. Recruiting staff unable to meet Graham's financial goals for the business would represent a failure, as would a major revolt by consultants in Tim's business. In the latter instance the professional services nature of the companies in this study may heighten the extent to which expert employees can assert their wishes in respect of changes to HRM practices owing to their importance to the business [66, 67, 78]. In both cases many decisions were arrived at informally, through owner manager prerogative or ad hoc negotiation with employees, not through the formalized systems and processes that the firms would present to external audiences.

4 Conclusion

In this chapter we have suggested that small firm HRM needs to be considered as distinct and researched on its own terms, attentive to ongoing, everyday practices. There is a risk with 'HRM' as discussed in the broad SME literature that it is from an inherently deficit position; it identifies what small firms are not doing or have not got in place [7]. Adapting conceptualizations of HRM processes from the large firm literature is an implicit way of holding an unhelpfully direct comparison between these very different kinds of firms. Small firms are different and, as such, our research has demonstrated one way in which their approaches to HRM practices can be understood on their own terms. Where existing studies often focus on debating the presence of HRM in smaller firms, we suggest that the debate should be less about whether or not some form of HRM exists in small firms but more about what functions it performs and how or when it is utilized [7].

While it is known from existing work that changes occur in small firms' HRM practices, our findings suggest something of how those changes come about. The changes in recruitment and selection and staff payment practices are consistent with notions of HRM in small firms being relatively ad hoc and informal and, importantly, our research demonstrates the ways in which such informality persists even when formal policies and practices are in place. The persistent informality of small firms influences how formality is practiced in the business and contributes to a diversity of experiences both between and within firms when it comes to changes in those practices.

The findings indicate that owner managers retain a degree of discretion with regards to certain HRM activities in their firms, albeit within the constraints of their operating context [24, 25, 49]. On this basis it may be possible to speculate that the particular balance of discretion and constraints in respect of practices come to shape the processes of how change occurs and HRM is enacted. For example, where an owner manager, or firm, possesses the discretion to implement changes and faces constraints in terms of resources, opposition or knowhow, short term changes to practices may occur. In contrast, where the discretion to alter practices is faced by no constraints, or constraints that can be readily overcome, enduring adaptation in HRM practices may be the result.

In the case of internal resistance and the constraints placed on owner managers by staff, this supports the notion of employees as not simply being passive recipients of owner manager control. Whilst employee influence is likely to be greater, more explicit and with more force behind it in professional service firms than lower skilled firms, different forms of negotiation and resistance may still limit the degree to which HRM practices can be adapted and embedded in the long term. It is also important to acknowledge the powerful forces exerting influence from outside many small firms, albeit forms of influence owner managers may accede to only superficially or in the short term. The forms such tensions and constraints take play important roles in the development and practice of HRM within small firms.

The potential for fluctuating, continuously changing formal practices or more short term trial and error approaches demonstrates limits on owner manager prerogative as well as something of the ways in which they seek to impose their desired forms of management over longer time frames as they seek to adapt HRM practices. These processes also suggest a risk of cross sectional studies of small firms that may miss these forms of change and the negotiations, constraints and tensions they reflect. Instead, such studies might inadvertently present relatively stable, strategic impressions of HRM practice and the ways in with processes change with firm size, formalization and in response to particular, rationally understood demands or challenges. Importantly, this study suggests forms of persistent informality that continue to influence ongoing, everyday practices after firms appear to have formalized and introduced written HRM policies.

Building on work that identifies that change occurs and the factors involved in promoting changes to create a picture of ad hoc informality, our findings suggest something of how those changes come about. The processes of how changes in recruitment and selection and staff payment practices occurred highlight that, while small firms may be largely reactive to their environment and ad hoc and informal in their approaches to HRM, the processes underpinning the forming and reforming of practices in use can differ when understood over the longer term and on an everyday basis. Understanding more about when and why these differences occur can advance understanding of HRM in small firms.

It can be suggested that recognizing how small firms' HRM practices alter in different ways may cause practitioners to reflect on the chances of HRM practices being successfully implemented and developed. The predominantly informal and ad hoc ways of organizing employment relationships, and other aspects of small firms, while not always ideal may suit the degrees of flexibility required by a firm's environment. Opting for particularly formal approaches to employment relationships may reduce this flexibility in favor of standardized processes that exist in tension with how a business operates. Judging when is the optimal time to increase degrees of formality, however, remains a matter of judgment [38] so practitioners should proceed with caution.

Researchers might consider further the differences in how changes to HRM practices in small firms occur and more generally question whether the practices they find currently in use are relatively stable, part of a broader process of trial and error or undergoing other processes of change. In this context, we suggest the importance of attending to persistent forms of informality for understanding even formalized firms with sophisticated HRM policies and practices. For both practitioners and researchers these initial findings suggest that there is value in deepening the characterization of small firms' HRM practices as ad hoc and informal.

References

1. Gill, J. (1985). *Factors affecting the survival and growth of the smaller company*. Aldershot: Gower Publishing Company Limited.
2. Wilkinson, A. (1999). Employment relations in SMEs. *Employee Relations, 21*(3), 206–217.
3. Ram, M., Edwards, P., Gilman, P., & Arrowsmith, J. (2001). The dynamics of informality: Employment relations in small firms and effects of regulatory change. *Work, Employment & Society, 15*(4), 845–861.
4. Kotey, B., & Sheridan, A. (2004). Changing HRM practices with firm growth. *Journal of Small Business and Enterprise Development, 11*(4), 474–485.
5. Dannreuther, C., & Perren, L. (2013). Uncertain states: The political construction of the small firm, the individualisation of risk and the financial crisis. *Capital & Class, 37*(1), 37–64.

6. Welsh, J. A., & White, J. F. (1981). A small business is not a little big business. *Harvard Business Review*, 18–32.
7. Taylor, S. (2005). The hunting of the snark: A critical analysis of human resource management discourses in relation to managing labour in smaller organizations. In S. Marlow, D. Patton, & R. Ram (Eds.), *Managing labour in small firms* (pp. 18–42). Abingdon: Routledge.
8. Torrès, O., & Julien, P.-A. (2005). Specificity and denaturing of small business. *International Small Business Journal, 23*(4), 355–377.
9. Curran, J. (2006). 'Specificity' and 'denaturing' the small business. *International Small Business Journal, 24*(2), 205–210.
10. Rainnie, A. (1989). *Industrial relations in small firms: small isn't beautiful*. London: Routledge.
11. Cassell, C., Nadin, S., Gray, M., & Clegg, C. (2002). Exploring human resource management practices in small and medium sized enterprises. *Personnel Review, 31*(6), 671–692.
12. Beaumont, P. B., Hunter, L. C., & Sinclair, D. (1996). Customer-supplier relations and the diffusion of employee relations changes. *Employee Relations, 18*(1), 9–19.
13. Kinnie, N., Purcell, J., Hutchinson, S., Terry, M., Collinson, M., & Scarbrough, H. (1999). Employment relations in SMEs: Market-driven or customer-shaped? *Employee Relations, 21*(3), 218–236.
14. Kinnie, N. J., Swart, J., & Purcell, J. (2005). Influences on the choice of HR system: The network organization perspective. *International Journal of Human Resource Management, 16*(6), 1004–1028.
15. Dundon, T., Grugulis, I., & Wilkinson, A. (1999). 'Looking out of the black-hole': Non-union relations in an SME. *Employee Relations, 21*(3), 251–266.
16. Westhead, P., & Storey, J. (1996). Management training and small firm performance: Why is the link so weak? *International Small Business Journal, 14*(4), 13–24.
17. Rodwell, J., & Shadur, M. (1997). What's size got to do with it? Implications for contemporary management practices in IT companies. *International Small Business Journal, 15*(2), 51–62.
18. Beaver, G., & Prince, C. (2004). Management, strategy and policy in the UK small business sector: a critical review. *Journal of Small Business and Enterprise Development, 11*(1), 34–49.
19. Marlow, S., & Patton, D. (1993). Managing the employment relationship in the smaller firm: Possibilities for human resource management. *International Small Business Journal, 11*(4), 57–64.
20. Jennings, P., & Beaver, G. (1997). The performance and competitive advantage of small firms: A management perspective. *International Small Business Journal, 15*(2), 63–75.
21. Gilman, M. W., & Edwards, P. K. (2008). Testing a framework of the organization of small firms: Fast-growth, high-tech SMEs. *International Small Business Journal, 26*(5), 531–558.
22. Mallett, O., & Wapshott, R. (2014). Informality and employment relationships in small firms: Humour, ambiguity and straight-talking. *British Journal of Management, 25*(1), 118–132.
23. Matlay, H. (1997). The paradox of training in the small business sector of the British economy. *Journal of Vocational Education and Training, 49*(4), 573–589.
24. Ram, M. (1994). *Managing to survive: Working lives in small firms*. Oxford: Blackwell.
25. Moule, C. (1998). Regulation of work in small firms: A view from the inside. *Work, Employment & Society, 12*(4), 635–653.
26. Ram, M. (1999). Managing autonomy: Employment relations in small professional service firms. *International Small Business Journal, 17*(2), 13–30.
27. Wapshott, R., & Mallett, O. (2013). The unspoken side of mutual adjustment: Understanding intersubjective negotiation in small professional service firms. *International Small Business Journal, 31*(8), 978–996.
28. Ram, M., & Holliday, R. (1993). Relative merits: Family culture and kinship in small firms. *Sociology, 27*(4), 48–629.
29. Jones, O. (2003). The persistence of autocratic management in small firms: TCS and organizational change. *International Journal of Entrepreneurial Behaviour & Research, 9*(6), 245–267.
30. Ram, M., & Edwards, P. (2010). Industrial relations in small firms. In T. Colling & M. Terry (Eds.), *Industrial relations: Theory and practice* (pp. 231–252). Chichester: Wiley.

31. Watson, T. J. (1995). Entrepreneurship and professional management: A fatal distinction. *International Small Business Journal, 13*(2), 34–46.
32. de Kok, J., & Uhlaner, L. M. (2001). Organization context and human resource management in the small firm. *Small Business Economics, 17*(4), 273–291.
33. Matlay, H. (1999). Employee relations in small firms: A micro-business perspective. *Employee Relations, 21*(3), 285–295.
34. Verreynne, M.-L., Parker, P., & Wilson, M. (2013). Employment systems in small firms: A multilevel analysis. *International Small Business Journal, 31*(4), 405–431.
35. Scott, M., Roberts, I., Holroyd, G., & Sawbridge, D. (1989). *Management and industrial relations in small firms*. London: Department of Employment.
36. Kotey, B., & Slade, P. (2005). Formal human resource management practices in small growing firms. *Journal of Small Business Management, 43*(1), 16–40.
37. Sheehan, M. (2013). Human resource management and performance: evidence from small and medium-sized firms. *International Small Business Journal, Available OnlineFirst.,*. doi:10.1177/0266242612465454.
38. Phelps, R., Adams, R., & Bessant, J. (2007). Life cycles of growing organizations: A review with implications for knowledge and learning. *International Journal of Management Reviews, 9*(1), 1–30.
39. Scholarios, D., & Lockyer, C. (1999). Recruiting and selecting professionals: Context, qualities and methods. *International Journal of Selection and Assessment, 7*(3), 142–156.
40. Carroll, M., Marchington, M., Earnshaw, J., & Taylor, S. (1999). Recruitment in small firms: Processes, methods, and problems. *Employee Relations, 21*(3), 236–250.
41. Harris, L. (2002). Small firm responses to employment regulation. *Journal of Small Business and Enterprise Development, 9*(3), 296–306.
42. Williamson, I. O. (2000). Employer legitimacy and recruitment success in small businesses. *Entrepreneurship Theory and Practice, 25*(Fall), 27–42.
43. Jameson, S. M. (2000). Recruitment and training in small firms. *Journal of European Industrial Training, 24*(1), 43–49.
44. Holliday, R. (1995). *Investigating small firms: Nice work?*. London: Routledge.
45. Marchington, M., Carroll, M., & Boxall, P. (2003). Labour scarcity and the survival of small firms: A resource-based view of the road haulage industry. *Human Resource Management Journal, 13*(4), 5–22.
46. Leung, A., Zhang, J., Wong, P. K., & Foo, M. D. (2006). The use of networks in human resource acquisition for entrepreneurial firms: Multiple 'fit' considerations. *Journal of Business Venturing, 21*(5), 664–686.
47. Heneman (III), H. G., & Berkley, R. A. (1999). Applicant attraction practices and outcomes among small businesses. *Journal of Small Business Management, 37*(1), 53–74.
48. Gilman, M., Edwards, P., Ram, M., & Arrowsmith, J. (2002). Pay determination in small firms in the UK: Contours of constrained choice. *Industrial Relations Journal, 33*(1), 52–67.
49. Arrowsmith, J., Gilman, M. W., Edwards, P., & Ram, M. (2003). The impact of the national minimum wage in small firms. *British Journal of Industrial Relations, 41*(3), 435–456.
50. Curran, J., Kitching, J., Abbott, B., & Mills, V. (1993). Employment and employment relations in the small service sector enterprise—A report. Kingston: Kingston University/Kingston Business School, ESRC Centre for Research on Small Service Sector Enterprises.
51. Forth, J., Bewley, H., & Bryson, A. R. (2006). Small and medium sized enterprises: findings from the 2004 Workplace Employment Relations Survey. London: Department of Trade and Industry; Economic and Social Research Council; Advisory, Conciliation and Arbitration Service; Policy Studies Institute.
52. Bacon, N., Ackers, P., Storey, J., & Coates, D. (1996). It's a small world: Managing human resources in small businesses. *International Journal of Human Resource Management, 7*(1), 83–100.
53. Jack, S., Hyman, J., & Osborne, F. (2006). Small entrepreneurial ventures culture, change and the impact on HRM: A critical review. *Human Resource Management Review, 16*(4), 456–466.

54. Rutherford, M. W., Buller, P. F., & McMullen, P. R. (2003). Human resource management problems over the life cycle of small to medium sized firms. *Human Resource Management, 42*(4), 321–335.
55. Redman, T., & Matthews, B. P. (1998). Service quality and human resource management: A review and research agenda. *Personnel Review, 27*(1), 57–77.
56. Mills, P. K., Chase, R. B., & Margulies, N. (1983). Motivating the client/employee system as a service production strategy. *Academy of Management Review, 8*(2), 301–310.
57. Doorewaard, H., & Meihuizen, H. E. (2000). Strategic performance options in professional service organizations. *Human Resource Management Journal, 10*(2), 39–57.
58. Morris, T., & Empson, L. (1998). Organization and expertise: An exploration of knowledge bases and the management of accounting and consulting firms. *Accounting, Organizations and Society, 23*(5/6), 609–624.
59. Goffee, R., & Scase, R. (1995). *Corporate realities: Dynamics of large and small organizations*. London: International Thomson Business Press.
60. Boxall, P. (2003). HR strategy and competitive advantage in the service sector. *Human Resource Management Journal, 13*(3), 5–20.
61. Behrends, T. (2007). Recruitment practices in small and medium size enterprises: An empirical study among knowledge-intensive professional service firms. *Management Revue, 18*(1), 55–74.
62. Scase, R. (1995). Employment relations in small firms. In P. Edwards (Ed.), *Industrial relations: Theory and practice* (pp. 569–595). Oxford: Blackwell.
63. Scase, R. (2005). Managerial strategies in small firms. In S. Marlow, D. Patton, & R. Ram (Eds.), *Managing labour in small firms* (pp. 66–82). Abingdon: Routledge.
64. Ram, M., & Edwards, P. (2003). Praising Caesar not burying him: What we know about employment relations in small firms. *Work, Employment & Society, 17*(4), 719–730.
65. Boxall, P., & Steeneveld, M. (1999). Human resource strategy and competitive advantage: A longitudinal study of engineering consultancies. *Journal of Management Studies, 36*(4), 443–482.
66. Tam, Y.-M. M., Korczynski, M., & Frenkel, S. J. (2002). Organizational and occupational commitment: knowledge workers in large corporations. *Journal of Management Studies, 39*(6), 775–801.
67. Swart, J., & Kinnie, N. (2003). Knowledge-intensive firms: The influence of the client on HR systems. *Human Resource Management Journal, 13*(3), 37–55.
68. Yin, R. K. (2003). *Case study research: Design and methods*. London: Sage.
69. Marlow, S. (2005). Introduction. In S. Marlow, D. Patton, & R. Ram (Eds.), *Managing labour in small firms* (pp. 1–17). Abingdon: Routledge.
70. Blackburn, R. (2005). Researching the employment relationship in small firms. In S. Marlow, D. Patton, & M. Ram (Eds.), *Managing labour in small firms* (pp. 43–65). Abingdon: Routledge.
71. Ram, M. (1999). Management by association Interpreting small firm-associate links in the business services sector. *Employee Relations, 21*(3), 267–284.
72. Grugulis, I., Dundon, T., & Wilkinson, A. (2000). Cultural control and the 'culture manager': Employment practices in a consultancy. *Work, Employment & Society, 14*(1), 97–116.
73. Miles, M. B., & Huberman, A. M. (1984). *Qualitative data analysis: A sourcebook of new methods*. Beverley Hills: Sage.
74. Miles, M. B., & Huberman, A. M. (1994). *Qualitative data analysis: An expanded sourcebook*. Thousand Oaks: Sage Publications Inc.
75. Galbraith, J. R., & Nathanson, D. A. (1978). *Strategy implementation: The role of structure and process*. St Paul: West Publishing Company.
76. Devanna, M. A., Fombrun, C. J., & Tichy, N. M. (1984). A framework for strategic human resource management. In C. J. Fombrun, N. M. Tichy, & M. A. Devanna (Eds.), *Strategic human resource management* (pp. 33–51). New York: Wiley.
77. Ram, M. (2000). Investors in People in small firms: Case study evidence from the business services sector. *Personnel Review, 29*(1), 69–91.
78. Swart, J. (2007). HRM and knowledge workers. In P. Boxall, J. Purcell, & P. M. Wright (Eds.), *The Oxford handbook of human resource management* (pp. 450–468). Oxford: Oxford University Press.

Effective Human Resource Practices in Family Businesses

Santiago Gutiérrez-Broncano, Mercedes Rubio-Andrés and Pedro Jiménez-Estévez

Abstract This chapter focuses on the need of effective practices of human resource management in the Family Business. Not so much literature we can obtain in this topic, but we have studied the main papers about this topic and we have obtained three main conclusions: the need to professionalize human resource practices in Family Business, the positive effect between human resource practices and firm's performance, and the differences about human resource practices in Family Business and Non-Family Business.

1 Introduction

In recent years, studies in family business (FB) in mainstream academic journals have been increasing [1, 2]. It reflects their economic significance in the countries and its significant role in employment creation, technological innovation and economic progress [3–5]. Traditionally, research in FB has focused on planning succession by younger generations, professionalizing through corporate governance, trying to discover how family involvement and interactions can generate unique resources and capabilities that contribute to create economic rent [6], or finding what uniqueness resources of family business improve the performance of the company [7]. Researches, scholars and professionals have been drawn to FB not only by its economic impact,

S. Gutiérrez-Broncano (✉) · P. Jiménez-Estévez
Department of Business Administration, Faculty of Social Sciences,
University of Castilla-La Mancha, Avda. Real Fábrica de Seda, s/n,
45600 Talavera de la Reina, Toledo, Spain
e-mail: santiago.gutierrez@uclm.es

M. Rubio-Andrés
Department of Business, School of Business and Economics,
San Pablo CEU University, C/Julián Romea 23, 1.08 B, 28003 Madrid, Spain

C. Machado and J. P. Davim (eds.), *Work Organization and Human Resource Management*, Management and Industrial Engineering,
DOI: 10.1007/978-3-319-06376-8_6,

but because some FBs have performance advantages and scholars have tried to explain these advantages by several reasons [3]. Recent evidence suggests that FB significantly outperform their rivals on assets and sales [8–11] or revenue higher for the first generation [10] and more firm longevity [12, 13].

To try to explain this situation, researches have used different theories. The main theoretical foundations have been based-resource view [14] but agency theory, stewardship theory, social exchange theory or even more recently socio-emotional theory [15] have been used too.

In addition, literature on Strategic Human Resource has demonstrated statistically significant relationships between a mix of Human Resource Practices (HRPs) (known as best practices in human resources) and firm profitability [16–18], turnover, productivity or corporate financial performance [19]. Effective HRPs are becoming increasingly important in the cotemporary knowledge-based economy [20] and companies must find ways to enhance their competitive advantage through effective use of their resources [21].

However, when we want to know about what HRPs are more effectiveness to family business, empirical research not only is scarce but lacks rigorous and quantitative analysis [22]. Little is known concerning human resource practices in FBs and their effects on performance [23].

Kidwell et al. [24] consider necessary increase attention in HRPs in FB because it is important in light of the rapid growth of family business. According to Kidwell et al. [24: 14] "HRPs and problems are key issues to the small and growing family business". By that, our study helps to address this lack of research by examining papers about HRPs in FB, which are in the top management journals includes in the most important and well-known databases: Web of Knowledge and Scopus.

So, we do not only make a review of the main works about this topic, but it is enough to know lacks and limitations. This work can be a good model to propose future research and it can be a good guide to FBs whose want improve its HRPs focus on attract, retain, motivate and develop Family Human Capital (FHC). We try to contribute to improve human resource management in family firms and avoid improper management of personnel in FBs.

2 Family Business Definition

Literature views FBs as heterogeneous and acknowledges that while some families can be assets to their firms and built unique resources, Known as *familiness* [25–29], other families could be characterized more as liabilities [30]. Familiness is described as "the idiosyncratic bundle of resource and capabilities resulting from the interaction of the family and business systems" [6: 45]. And *familiness* is often used as a unique element that cans differentiate family and non-family business [25]. Family members simultaneously participate in both business and family relationships in their personal and professional lives. These relationships increase their complexity and create a unique context for human capital compared to non-family business. Traditional definitions of FBs have been focus on some combination of the components of a family's involvement in the business: ownership, governance, management, and trans-generational succession [31]. But *familiness* refers to the combination on existing stocks of social, human, financial and physical capital resources in a firm resulting from interactions between family and business systems [6, 29].

Others scholars define family business by its essence and they evict that firms with the same extent of family involvement may or may not consider themselves FB or it may change over time [28]. Its essence can be a family's influence over the strategic direction of a company [32] or the intention of the family to keep control [33] or family firm behavior [31] or even unique inseparable, synergetic resources and capabilities arising from family involvement and interactions [6].

A more practical definition is what define family business as "a family member is chief executive, they are at least two generations of family control (and) a minimum of five percent of voting stock in held by the family or trust interested associated with it" [34, 30]. Recently, Klein et al. [35] have developed and validated a scale (F-PEC) to measure this involvement as a continuous variable measuring three dimensions: power, experience and culture. The dichotomous in the concept of FB can create findings and loss of important information [11]. Nevertheless, definition of FB stays open to debate [28].

But, is it possible that FBs are not different from non-FBs? From strategy point of view if some aspects of FBs are different to non-FBs and its aspects affect to performance, there is no doubt about both are different, because FBs have performance advantage as we saw before. The question now is, why?

3 The Importance of Family Human Capital

Scholars have been paid increasing attention recently to the role of human resources factors in influencing firm competitiveness. While tangible assets are easily imitable and thus unlikely to be a source of sustainable competitive advantage [14], human assets are often hard to imitate [36]. So, competent people are the key to success and offer organizations their only sustainable competitive advantage [37].

Human capital refers to knowledge, skills and technical abilities, but in the context of family firms we have to add emotional strength or carrying capacity and intellectual capital of family and non-family members. FBs are often likely to create a sustainable competitive advantage over non-FBs and it is thanks to the attitudes of family members and not only by family members' knowledge, skills and abilities [3]. Literature distinguishes between human capital and Family Human Capital (FHC) because the last one adds a dimension around attitudes and motivation that Dawson [3] recently has named "heart". Some aspects like extraordinary commitment [38, 39], long term orientation, alignment between the human capital pool and the strategic goals of the firm, warm, friendly and intimate relationships and the potential tacit knowledge, influence to make decision in FB. For example, FBs have a stronger tendency to keep their employees even in times of economic crisis [40] and altogether this is likely to lead to increased trust or feeling of job security and commitment [41].

FHC makes that although FBs often do not possess a superior pool of human capital, in terms of knowledge, skills and abilities, they are able to achieve competitive advantage [36, 42]. Generally, FBs have a limited pool of potential recruits and may not be able to supply the firm with enough talented employees to manage the key operations. FB owners are often criticized for employing children who are less able, committed or ethical than leader expected [43]. But although it is true, it has a special importance in firms that require highly specialized knowledge of technology and markets or firms that are very large and complex [44] not in all of them. In contrast, family members generally have a high alignment with business because they was seen themselves as the organization's goals and values [45]. The closeness of the family/business relationship creates a unique context for human capital (Both positive and negative), compared to non-FBs [46].

Literature has identified several motivational mechanisms that induce individuals to behave in line with organizational goals and which are related with human resource practices. For instance, appropriate reward system, flexible job design and strong socialization system, are traditional ways to improve the alignment between organizations and their members [36, 47].

Other authors found another types of causes like the family name. They argue that if the family name is "on the building", family members will naturally be more motivated and committed to the business [48, 49].

According to Dyer [44] family connections inspire loyalty and family members are therefore willing to work long hours –often without compensation- and be highly flexible in their work roles and assignment in order to help the firm succeed. Family members have often socialized at a very early age to understand the nature of the business, its costumers and its competitors.

In contrast, if we accept nepotism as norm in family business, family members who are incompetents will be placed in key positions and family business will not be utilizing the best management talent [44]. If owners violate significant organizational norms threatens the well-being of the organization and/or its members [50]. It can include free riding, lying, insubordination, sabotage, abuse of privileges, harassment, theft, aggression, lack of regard for cost control and drug and alcohol abuse [51]. Repercussions of it are lawsuits productivity losses, higher expenses and loss of reputation.

About that, Stalk and Foley [52: 26] talk about three traps of fb: firstly, "there is always a place for you here"; secondly, "the business can't grown fast enough to support every one" and thirdly, "family members remain in silos according to bloodlines".

To analyze these issues, researchers have used different theories more or less focus on FBs. Traditionally, researches have used agency theory and resource-based view to explain the differences between family and non-family firms [20]. Using agency theory, less professional HRP are required to align interest of managers and employees when both belong to the same family [20]. If we use resourced-based view we have use the concept of *faminess*. According to it human capital can be higher in FBs than non-FBs.

Contingency approaches assume that HRP are instituted to elicit and control employee behaviors in ways that will be most effective for the organization depending upon the characteristics, needs and circumstances of the firm [53].

In contrast, the institutional perspective argue that organizations select, adapt and retain many structures and practices for normative reasons due to the fact such practices have obtained legitimacy within an industry usually due to external forces [54, 55].

4 Human Resources Practices in Family Business

As it was said before, very little literature examines HRPs at family-controlled [56] and little is known concerning HRPs in FBs and their effects on performance [23]. Despite this, literature in Family Business suggests that FB have difficulty managing their HR, especially when it concerns family members [57]. It is also extremely difficult for a CEO who is the father, mother or uncle to practice HRPs objectively with family members [58]. Some authors recommend that family business HRP may require greater clarity in the areas separating ownership and management [59]. Ward [60] emphasizes strongly that effective management in family business requires the establishment of structures and processes which link management and ownership. These can be especially helpful in the area of human resource management to improve the practice of selection, compensation and equity, appraisals, training and development and promotion [58].

The main problem caused by the overlap between the family and the business in the area of human resources were set out wisely by Lansberg in 1983. While family is guided by implicit affective principles that focus on each person's attention on the needs and long-term well-being of the other, firm is regulated by economic principles.

Based on Institutional theory for show the negative effects that appear when norms and principles that operate in family, operate in business too, Lansberg [59] proposed to develop effective procedures for helpful in the area of human resource management to improve the practices of selection, compensation on equity, appraisals, training and development and promotion in FBs, the separation of management and ownership [59]. According to this, we find more literature and consensus about the need to professionalize human resources management within FBs because FBs are less likely to have professional HRPs [7, 20, 58].

One of the first works that treated this subject in FB area was en 1997, where Aldrich and Langton [61] found negative relationship between the number of family members and formal HRP. Afterwards, Reid et al. [62] found that FBs in Northern Ireland lagged non-FBs implementing HRM policies and practices but found indication that news generations of family ownerships would like transition toward more formal HRP.

Table 1 Human resource practices in FBs and non-FBs

Higher for family business	Higher for non-family business
Level of turnover	Labor cost
Use flat rate and individual bonus pay mechanism	Negotiation with trade unions
	Having a personnel or human resource management department
	Use reference as a selection method
	Use of appraisals systems
	Use merit-performance related pay as a pay reward mechanism for managers
	Cost of their annual salaries and wage bill on training
	Reported systematically analyzing employee training needs
Non differences between FB and non-FBs	
Less than 50 % or both family and non-family reported that the head of human resource management did not have a place on the board	
Training and development are considered the major challenge for HRPs in their businesses over the next 5 years	

Source Adapted Reid and Adams [58]

A year after, Reid and Adams [58] found that FB are less likely to have a Human Resource Department and non-FBs was more likely to have a Human Resource Manager. In the same way, non-FBs was more likely to use appraisals systems, set merit/performance related pay for managers and keep detailed HRM records [58]. Similar results were found by Kok [20] in a sample of approximately 700 small to medium-sized firms. Over more, research has indicated that FBs are slower to adopt HRP when we compared to their non-family counterparts [7].

Recently, one of the last studies in this topic has been carry out in 205 FBs in the People's Republic of China and it shows as the greater size of the firm the more formalized its HRP [63].

If we focus on the differences between HRP in FBs and non-FBs, little empirical evidence we can find, in spite of HRPs appear to be one area where FBs have adopted significantly different practices from their non-family counterparts. Some differences between HRPs in FBs and non-FB in SMEs recently found by Reid and Adams [58], we can see in the Table 1.

In a recent study in Spanish companies, Carrasco and Sánchez [64] found similar results about employee compensation. Pay level is lower in FBs than in both non-FBs and professionally FBs and about regarding pay mix, employees in professionally managed receive the highest proportion of variable pay in their compensation packages compared to FBs and non-FBs [64].

Finally when we study the relationship between HRPs and performance we found little consensus among researchers regarding the specific policies to be included in configurations of high performance human resource practices [65–70]. However various studies have suggested that these may include rigorous selection procedures, internal merit-based promotions, cross-trained teams, high levels of training, information sharing, skill-based pay and group-based rewards [71–75]. Sets of such practices have been named "high performance work practices" [19, 70], "strategic human resource practices" [76, 77], "high involvement practices" [78] or "high commitment practices" [79, 80]. Such high performance human resource practices aim to enhance employee engagement across the company [81].

In recent years scholars have tried to identify human resource practices, or bundles of human resource practices, which bring competitive advantage to companies. At first, practices concerning information, training, remuneration and empowerment [78, 82] were selected as best practices to improve organizational performance. Then, other scholars added more practices such as extensive training and high salaries [75] or selective recruitment, appraisals systems and job security [83].

And although much of the literature about strategic human resource argues that there is considerable evidence that these practices are associated with high organizational performance [17, 19, 73, 75, 84–88], not so in the case of FBs, where little is concerning about HRPs in FBs and their effect on performance [23] and heterogeneous methodology has been used.

Within the FB literature there is a positive association between HRPs like formal employee reviews, set compensation plans, written employee manuals, and use of written job description, and gross firm revenues [41].

A study realized to 168 family-owned small and medium enterprises (SMEs) examine the consequences of five HRPs on sales growth performance. Training and development, performance appraisals, recruitment package, maintaining morale and setting competitive compensation levels, do in fact have a positive impact on performance in FBs [56]. According to it, Kidwell and Fish [7] examine the prevalence of high-performance human resource systems in Australian FBs and found that increase emphasis on employee involvement and motivation was positively related to operational and financial success.

Similar results were obtained later in a study focus on Australian and American FBs wineries [89]. Despite this, not all HRPs were related with performance. For instance, employee involvement, training, and HR systems and alignment were perceived as an overhead expense in these companies and only motivational practices and leadership were positively correlated with sales revenue, ROI, and profitability [89]. And when we analyzed public firms, ownership and firm performance are not associated, but both are related when high performance human resources practices are implemented in the company [90].

5 Conclusions

One of the most important resources that possess family businesses is family human capital [46]. Knowledge, skills and abilities of individual family members are not sufficient to create value for an organization, unless they are used through individual behaviors [91]. Thus, a family business needs to manage an alignment of interests in order to create a committee workforce [92].

This is very important because family business does not own the human capital, the family members do. Family businesses need higher levels of human capital and achieve a superior alignment between individuals and organizational [42] in order to achieve competitive advantage.

To obtain it, FBs have to implement an effective system of Human Resource Practices. Literature show us how family businesses need professionalize its human resource practices to improve its. But if family businesses want maintain and increase their family human capital and so get differentiate from non-family businesses, they have to focus on some practices of human resource and non necessary the same of the non-family business.

In conclusion, we would like to point to some key research questions which we believe need to be addressed:

- We have seen like FBs and non-FBs use different practices of Human Resource, nevertheless FB has more Family human Capital and obtain higher performance, can we propose a different way of manage its human resources, more different or similar than non-FB?
- Knowing the heterogeneous FBs that exist, can we speak about the most effective HRPs in FBs or will it depend on the type of FBs?
- Will this knowledge be useful for managing FBs and improving their competitive position?
- What will we achieve with it? More commitment among employees, better relationships between family and non-family members, … In general, more Human Capital or more Family Human Capital?

References

1. Stewart, A., & Miner, A. S. (2011). The prospects for family business in research universities. *Journal of Family Business Strategy, 2*, 3–14.
2. Sharma, P., Hoy, F., Astrachan, J. H., & Koiranen, M. (2007). The practice-driven evolution of family business education. *Journal of Business Research, 60*, 1012–1021.
3. Dawson, A. (2012). Human capital in Family Businesses: Focusing on the individual level. *Journal of Business Strategy, 3*, 3–11.
4. Zahra, S. A., Hayton, J. C., Neubaum, D. O., Dibrell, C., & Craig, J. (2008). Culture of family commitment and strategic flexibility: The moderating effect of stewardship. *Entrepreneurship Theory and Practice, 32*(6), 1035–1054.
5. Zahra, S. A. (2005). Entrepreneurial risk taking in family firms. *Family Business Review, 18*(1), 23–40.
6. Habbershon, T. G., Williams, M., & MacMillan, I. C. (2003). A unified systems perspective of family firm performance. *Journal of Business Venturing, 18*, 451–465.
7. Kidwell, R. L., & Fish, A. J. (2007). High-performance human resource practices in Australian family businesses: Preliminary evidence from the wine industry. *International Entrepreneurship Management Journal, 3*, 1–14.
8. Anderson, R., & Reeb, D. (2003). Founding family ownership and firm performance: Evidence from the S&P 500. *Journal of Finance, LVIII*(3), 1308–1328.
9. Anderson, R. C., Mansi, S., & Reeb, D. (2003). Founding family ownership and the agency cost of debt. *Journal of Financial Economics, 68*, 263–285.

10. Weber, J., et al. (2003). Family inc. *Business Week, 10*, 100–114.
11. McConaughy, D., Matthews, C., & Fialco, A. (2001). Founding family controlled firms: Performance, risk and value. *Journal of Small Business Management, 39*, 31–49.
12. Mackie, R. (2001). Family ownership and business survival. *Business History, 43*, 1–32.
13. De Geus, A. (1997). *The living company*. Boston: Harvard Business School Press.
14. Barney, J. B. (1991). Firm resources and sustained competitive advantage. *Journal of Management, 17*, 99–120.
15. Berrone, P., Cruz, C., & Gómez-Mejía, L. R. (2012). Socioemotional wealth in family firms: Theoretical dimensions, assessment approaches, and agenda for future research. *Family Business Review, 25*(3), 258–279.
16. Camps, J., & Luna-Arocas, R. (2012). A matter of learning: How human resources affect organizational performance. *British Journal of Management, 23*, 1–21.
17. Wright, P. M., Gardner, T. M., Moynihan, L. M., & Allen, M. R. (2005). The relationship between HR practices and firm performance: Examining causal order. *Personnel Psychology, 58*, 409–446.
18. Delery, J. E., & Doty, D. H. (1996). Modes of theorizing in strategic human resource management: Tests of universalistic, contingency, and configurational performance predictions. *Academy of Management Journal, 39*(4), 802–835.
19. Huselid, M. A. (1995). The impact of human resource management practices on turnover, productivity, and corporate financial performance. *Academy of Management Journal, 38*, 635–672.
20. Kok, J. M. P., Uhlaner, L. M., & Thurik, A. R. (2006). Professional HRM practices in family owned-managed enterprises. *Journal of Small Business Management, 44*(3), 441–460.
21. Senge, P. M., & Fulmer, R. M. (1993). Simulations, systems thinking and anticipatory learning. *Journal of Management Development, 12*(6), 21–34.
22. Heneman, R. L., Tansky, J. W., & Camp, S. M. (2000). Human resource management practices in small and medium-sized enterprises: Unanswered questions and future research. *Entrepreneurship Theory and Practice, 25*, 1–11.
23. Reid, R. S., Morrow, T., Kelly, B., & McCartan, P. (2002). People management in SMEs: An analysis of human resource strategies in family and non-family businesses. *Journal of Small Business and Enterprise Development, 9*(3), 245–259.
24. Kidwell, R. E., Hoy, F., & Ibarreche, S. (2012). "Ethnic" family business or just family business? Human resource practices in the ethnic family firm. *Journal of Family Business Strategy, 3*, 12–17.
25. Pearson, A. W., Carr, J. C., & Shaw, J. C. (2008). Toward a theory of familiness: A social capital perspective. *Entrepreneurship Theory and Practice, 32*(6), 949–969.
26. Rutherford, M. W., Kuratko, D. F., & Holt, D. T. (2008). Examining the link between "Familiness" and performance: Can the F-Pec untangle the family business theory jungle? *Entrepreneurship Theory and Practice, 32*(6), 1089–1110.
27. Naldi, L., Nordqvist, M., Sjoberg, K., & Wiklund, J. (2007). Entrepreneurial orientation, risk taking, and performance in family firms. *Family Business Review, 20*(1), 33–47.
28. Chrisman, J. J., Chua, J. H., & Sharma, P. (2005). Trends and directions in the development of a strategic management theory of the family firm. *Entrepreneurship Theory and Practice, 29*(5), 555–575.
29. Habbershon, T. G., & Williams, M. (1999). A resource-based framework for assessing the strategic advantage of family firms. *Family Business Review, 12*, 1–25.
30. Kellermanns, F. W., Eddleston, K. A., Sarathy, R., & Murphy, F. (2012). Innovativeness in family firms: a family influence perspective. *Small Business Economics, 38*, 85–101.
31. Chua, J. H., Chrisman, J. J., & Sharma, P. (1999). Defining the family business by behavior. *Entrepreneurship Theory and Practice, 23*, 19–39.
32. Davis, J. A., & Tagiuri, R. (1989). The influence of life-stage on father–son work relationships in family companies. *Family Business Review, 2*, 47–74.
33. Litz, R. (1995). The family business: Toward definitional clarity. In D. P. Moore (Ed.), *Proceedings of the academy of management* (pp. 100–104). Canada: Vancouver.

34. Colli, A., Fernandez-Perez, P., & Rose, M. (2003). National determinants of family firm development: Family firms in Britain, Spain and Italy in the 19th and 20th centuries. *Enterprise & Society, 4*(1), 28–65.
35. Klein, S. B., Astrachan, J. H., & Smyrnios, K. X. (2005). The F-PEC scale of family influence: construction, validation, and further implication for theory. *Entrepreneurship Theory and Practice, 29*(3), 321–339.
36. Gottschalg, O., & Zollo, M. (2007). Interest alignment and competitive advantage. *Academy of Management Review, 32*(2), 418–437.
37. Pickett, L. (1998). Competencies and managerial effectiveness: putting competencies to work. *Public Personnel Management, 27*(1), 103–116.
38. Horton, T. P. (1986). Managing in a family way. *Management Review, 75*(2), 3.
39. Donnelley, R. (1964). The family business. *Harvard Business Review, 42*(2), 93–105.
40. Lee, J. (2006). Impact of family relationships on attitudes of the second generation in family business. *Family Business Review, 19*(3), 175–191.
41. Astrachan, J. H., & Kolenko, T. A. (1994). A neglected factor explaining family business success: human resource practices. *Family Business Review, 7*(3), 251–262.
42. Wright, P. M., Dunford, B. B., & Snell, S. A. (2001). Human resources and the resource based view of the firm. *Journal of Management, 27*, 701–721.
43. Chrisman, J. J., Chua, J. H., & Litz, R. A. (2004). Comparing the agency costs of family and non-family firms: Conceptual issues and exploratory evidence. *Entrepreneurship Theory and Practice, 28*, 335–354.
44. Dyer, W. G. (2006). Examining the "family effect" on firm performance. *Family Business Review, 19*, 253–273.
45. Meyer, J. P., & Herscovitch, L. (2001). Toward a general model of commitment. *Human Resource Management Review, 11*(3), 299–326.
46. Sirmon, D. G., & Hitt, M. A. (2003). Managing resources: Linking unique resources, management and wealth creation in family firms. *Entrepreneurship Theory and Practice, 27*(4), 339–358.
47. Lindenberg, S. (2001). Intrinsic motivation in a new light. *Kyklos, 54*, 317–342.
48. Ward, J. L. (1988). The special role of strategic planning for family businesses. *Family Business Review, 1*(2), 105–118.
49. Rosenblatt, P. C., de Mik, L., Anderson, R. M., & Johnson, P. A. (1985). *The family in business*. San Francisco: Jossey-Bass.
50. Robbinson, S. L., & Bennett, R. J. (1997). Workplace deviance: Its definition, its manifestations, and its causes. *Research on Negotiation in Organizations, 6*, 3–27.
51. Eddleston, K. A., & Kidwell, R. E. (2012). Parent-child relationships: Planting the seeds of deviant behavior in the family firm. *Entrepreneurship Theory and Practice, 36*(2), 369–386.
52. Stalk, G., & Foley, H. (2012). Avoid the traps that can destroy family businesses. *Harvard Business Review*, 90 (1/2), 25–27.
53. Ericksen, J., & Dyer, L. (2005). Toward a strategic human resource management model of high reliability organization performance. *International Journal of Human Resource Management, 16*, 907–925.
54. Paauwe, J., & Boselie, P. (2005). Best practices in spite of performance: Just a matter of imitation. *International Journal of Human Resource Management, 16*, 987–1003.
55. Wright, P. M., & McMahan, G. C. (1992). Theoretical perspectives for strategic human resource Management. *Journal of Management, 18*, 295–320.
56. Carlson, D. S., Upton, N., & Seaman, S. (2006). The impact of human resource practices and compensation design on performance: An analysis of family-owned SMEs. *Journal of Small Business Management, 44*(4), 531–543.
57. King, S. W., Solomon, G. T., & Fernald, L. W. (2001). Issues in growing a family business: A strategic human resource model. *Journal of Small Business Management, 39*(1), 3–13.
58. Reid, R. S., & Adams, J. S. (2001). Human resource management: A survey of practices within family and non-family firms. *Journal of European Industrial Training, 25*(6), 310–320.

59. Lansberg, I. (1983). Managing human resources in family firms: The problem of institutional overlap. *Organizational Dynamics, 12*(1), 39–46.
60. Ward, J. L. (1987). *Keeping the family business healthy*. San Francisco: Jossey-Bass.
61. Aldrich, H., & Langton, N. (1997). Human resource management practices and organizational life cycles, In P. D. Reynolds, W. D. Bygrave, N. M. Carter, P. Davidsson, W. B. Gartner, C. M. Mason, and P. P. McDougall (Eds.), *Frontiers of entrepreneurship research.Babson college center for entrepreneurship* (pp. 349–357) Wellesley, MA.
62. Reid, R., Morrow, T., Kelly, B., Adams, J., & McCartan, P. (2000). Human resource management practices in SME's: A comparative analysis of family and non-family businesses. *IBAR-Journal of the Irish Academy of Management, 21*(2), 157–181.
63. Kim, Y., & Gao, F. Y. (2010). An empirical study of human resource management practices in family firms in China. *The International Journal of Human Resource Management, 21*(12), 2095–2119.
64. Carrasco-Hernandez, A., & Sánchez-Marín, G. (2007). The determinants of employee compensation in family firms: Empirical evidence. *Family Business Review, 20*, 215–228.
65. Collins, C. J., & Smith, K. G. (2006). Knowledge exchange and combination: The role of human resource practices in the performance of high-technology firms. *Academy of Management Journal, 49*, 544–560.
66. Datta, D. K., Guthrie, J. P., & Wright, P. M. (2005). Human resource management and labor productivity: Does industry matter. *Academy of Management Journal, 48*, 135–145.
67. Delery, J. E., & Shaw, D. (2001). The strategic management of people in work organizations: Review, synthesis and extension. In G. R. Ferris (Ed.), *Research in personnel and human resource management*, 20, (pp. 165–197). Stamford: JAI Press.
68. Ferris, G. R., Arthur, M. M., Berkson, H. M., Harrel-Cook, G., & Fink, G. (1998). Toward a social context theory of the human resource management-organizational effectiveness relationship. *Human Resource Management Review, 8*, 235–264.
69. Becker, B. E., & Gerhart, B. (1996). The impact of HRM organizational performance: Progress and prospects. *Academy of Management Journal, 39*, 779–801.
70. Delaney, J. T., & Huselid, M. A. (1996). The impact of human resource management practices on perceptions of organizational performance. *Academy of Management Journal, 39*, 949–969.
71. Qiao, K., Khilji, S., & Wang, X. (2009). High-performance work systems, organizational commitment, and the role of demographic features in the People's Republic of China. *The International Journal of Human Resource Management, 20*(11), 2311–2330.
72. Sun, L. Y., Aryee, S., & Law, K. S. (2007). High-performance human resource practices, citizenship behavior, and organizational performance: a relational perspective. *Academy of Management Journal, 50*(3), 558–577.
73. Guthrie, J. P. (2001). High involvement work practices, turnover and productivity: Evidence from New Zealand. *Academy of Management Journal, 44*, 180–190.
74. Levine, D. (1995). *Reinventing the workplace: How business and employers can both win*. Washington, DC: Brookings Institution.
75. Arthur, J. (1994). Effects of human resource systems on manufacturing performance and turnover. *Academy of Management Journal, 37*, 670–687.
76. Hiltrop, J. M. (1999). The quest for the best: Human resource practices to attract and retain talent. *European Management Journal, 17*(4), 422–430.
77. Huselid, M. A., Jackson, S. E., & Schuler, R. S. (1997). Technical and strategic human resource management effectiveness as determinants of firm performance. *Academy of Management Journal, 40*(1), 171–188.
78. Lawler, E. (1986). *High-involvement management*. San Francisco: Jossey-Bass.
79. Kwon, K., Bae, J., & Lawler, J. J. (2010). High commitment HR practices and top performers impacts on organizational commitment. *Management International Review, 50*, 57–80.
80. Meyer, J. P., & Smith, C. A. (2000). HRM practices and organizational commitment: Test of a mediation model. *Canadian Journal of Administrative Sciences Revue canadienne des sciences de l'administration, 17*(4), 319–331.

81. Guerrero, S., & Barraud-Didier, V. (2004). High-involvement practices and performance of French firms. *The International Journal of Human Resource Management, 15*(8), 1408–1423.
82. Leana, C., & Florkowski, G. (1992). Employee involvement programs: Integrating psychological theory and management practice. *Research in Personnel and Human resource management, 10*, 233–270.
83. Wood, S., & De Menezes, L. (1998). High commitment management in the UK: Evidence from the workplace industrial relations survey, and employers' manpower and skills practices survey. *Human Relations, 51*(4), 485–515.
84. Bartel, A. P. (2004). Human resource management and organizational performance: Evidence from retail banking. *Industrial and Labour Relations Review, 57*, 181–203.
85. Batt, R. (2002). Managing customer services: Human resource practices, quit rates, and sales growth. *Academy of Management Journal, 45*, 587–597.
86. Appelbaum, E., Bailey, T., Berg, P., & Kalleberg, A. (2000). *Manufacturing advantage: Why high-performance work systems pay off*. Ithaca: ILR Press.
87. Bae, J., & Lawler, J. J. (2000). Organizational and HRM strategies in Korea: Impact on firm performance in an emerging economy. *Academy of Management Journal, 43*, 502–517.
88. MacDuffie, J. P. (1995). Human resource bundles and manufacturing performance. *Industrial and Labor Relations Review, 48*, 197–221.
89. Thach, L., & Kidwell, R. E. (2009). HR practices in US and Australian family wineries: cultural contrasts and performance impact. *International Entrepreneurship Management Journal, 5*, 219–240.
90. Tsao, C. W., Chen, S. J., Lin, C. S., & Hyde, W. (2009). Founding-family ownership and firm performance: The role of high-performance work systems. *Family Business Review, 22*, 319–332.
91. Colvin, A. J. S., & Boswell, W. R. (2007). The problem of action and interest alignment: Beyond job requirements and incentive compensation. *Human Resource Management Review, 17*(1), 38–51.
92. Boxall, P. F. (1996). The strategic HRM debate and the resource-based view of the firm. *Human Resource Management Journal, 6*(3), 59–75.

The Gender of the Entrepreneur and the Politics of Entrepreneurship for Women: An Exploratory Study

Emília Fernandes and Regina Leite

Abstract Based on interviews conducted with Portuguese Institutional Business Associations located in the northern region, we propose firstly to critically analyze how and in what way such organizations that formally represent Portuguese entrepreneurs, at national or regional level, use certain gender logics of inclusion/exclusion to constitute the normative gendered subject of entrepreneurship. Secondly, we intend to further analyze how these organizations explain, justify and legitimize policies of positive gender discrimination and affirmative action that they promote to increase participation of women in such context. More specifically, we aim to reflect upon the tension that was found in the level of reflexivity related with a discourse of gender equality and how it contributed to rethinking women's condition in the entrepreneurial practice.

1 Introduction

In recent years, entrepreneurship has assumed a particular relevance in the Portuguese political discourse as a way to solve the problem of unemployment and "shorten" the participation of the Welfare State. In addition to this national interest, policy guidelines and investments undertaken by the European Union have been made in consulting and training in entrepreneurship. These policies have gained higher visibility in the public opinion when they focused exclusively on social groups considered to be discriminated in the context of business, that gave women the status of "minority statistics" [1]. Such policies were intended to fight the masculinization of these contexts and facilitate the introduction of women as legitimate entrepreneurs. The practice of entrepreneurship and its subjects, as in other work places,

E. Fernandes (✉) · R. Leite
Department of Management, School of Economics and Management,
University of Minho, Campus Gualtar, 4710-057 Braga, Portugal
e-mail: mifernandes@eeg.uminho.pt

C. Machado and J. P. Davim (eds.), *Work Organization and Human Resource Management*, Management and Industrial Engineering,
DOI: 10.1007/978-3-319-06376-8_7,

has been social and culturally constructed around a certain ideal of masculinity that serves to define what the practice of entrepreneurship is and who can be an entrepreneur. We can call the masculine normative discourse "mentality of entrepreneur", as named by Bruni et al. [2], which is related to a symbolic masculinity that emphasizes the ambitious individual, the ability of taking risks and enterprise profit. A discourse through which women are also constituted as the other [3, 4], i.e. the marginal gender subject that is always able to fulfil such personal and individual profile and that, at best, can only aim to complement the authentic male entrepreneur.

The need to symbolically link entrepreneurship to masculinity and the women's need to be associated with this symbolism is somehow very present in some training and consulting measures launched by Portuguese governmental institutions to foster entrepreneurship among women. An example that illustrates this relationship and the need for such masculinity to legitimize the entrepreneurship activity of women is the image used by the Portuguese government to announce an entrepreneurship training programme directed at women only. Such announcement was widely publicized in the beginning of 2009 at several Portuguese journals, and presented a smiling young woman looking at the horizon. An unreachable horizon to the viewer since it fell outside the image boundaries. We assume that the position of women's body and face is intended to transmit the idea that she believes in her entrepreneurship future. This image is easy to understand since the target audience of this entrepreneurship training are women that might be interested in engaging in an entrepreneur "adventure". We can also see in the picture a masculine figure, even though blurred and placed second. This male figure that embraces the young woman is sharing with her the same smile and the same look at the horizon. So, we can question why the need to put such male figure in an announcement that aims to promote an entrepreneurship training program exclusively for women? It seems that the "feminine entrepreneurship[1]" can only be thought or turned possible with the masculine consent or participation in such entrepreneurial project. We can also presume that the masculine presence can in some way legitimize and validate this idea of "feminine entrepreneurship" since men are also invited to be present, although in a second place. Besides making the entrepreneurial feminine project viable, the masculine presence can also be regarded as a warranty in overcoming the gendered condition if the project remained circumscribed to women.[2]

This illustration based on the government announcement of training programs exclusively for women was used to draw attention to and reflect upon the institutional discourses of gender that serve to explain, justify and legitimize policies of positive gender discrimination and affirmative action in entrepreneurship. With this aim in mind, we hope to give an account of the logics of power produced by such discourses that end up constituting the normative practices and subjects of entrepreneurship.

[1] The title of the announcement.

[2] We could also make many considerations about the discourses in which such picture can be inscribed, related to ethnicity, age and heteronormativity. Nevertheless, such identity categories are not explored in this chapter.

Following the European Union guidelines, Portugal has adopted some measures aimed at promoting the participation of women in entrepreneurship, namely some affirmative actions materialized in training programs for women. The Institutional Business Associations emerge as privileged interlocutors between Portugal and the European Union in the implementation of policies intended to promote equal opportunities between genders in entrepreneurship. This is one of the reasons why we have decided to contact six of the major and more important Institutional Business Associations of the north of Portugal, with the purpose of exploring the system of meanings associated with the ideal figure of the entrepreneur subject and their signification and positioning regarding training programs of entrepreneurship exclusively for women. Using semi-structured interviews we explore the reflections of this study's participants about the economic and entrepreneurial tissue focusing on the rhetorical discourses of these organizations in order to identify differences in terms of the entrepreneurs that they must represent (businesswomen, young entrepreneurs, small and medium entrepreneurs, entrepreneurs at regional or national level) and their level of knowledge, receptivity and implication in women's entrepreneurship training programs.

The chapter begins with a brief reflection on the gender of entrepreneurship and the construction of women as the "other" in such organizational and business activity. It proceeds with a brief presentation of the methodology, namely the interactional process in the relation between the participants chosen for this study and the researcher—a female researcher—that presented the study to them and conducted the interviews. At the end of the chapter we will present the results of the discourse analysis of interviews, and reflect upon the tensions that we have found between a discourse of gender equality and a discourse of gender difference in the explanation and justification of the affirmative policies, and how this tension contributes to reproduce or subvert the positionality of "other" that has been reserved to the female entrepreneurs in the entrepreneurial practice.

2 The Masculinity of Entrepreneurship Practice and Politics of Gender Equality in Entrepreneurship

Existing knowledge about the practice of entrepreneurship has been dominated by positivists and normative theoretical models that tend to confirm the cultural fictions developed around such practice as an elitist and exceptional practice that only few (men) are able to perform [4–6]. Cultural fictions that are frequently reproduced by acritical research and education and that inscribes entrepreneurship in the metanarrative of economic progress and modernity curricula [7]. As Bruni et al. [2] remember, the triptychs risk-innovation-profit are determinants of the ideal construction of the practice of entrepreneurship and constitute a disciplinary power in everyday life of the work and business contexts. Nevertheless such discourse is able to fulfil all the diversity that can be found in the entrepreneurial act and its subjects [8, 9].

From a critical point of view, the act of entrepreneurship can paradoxically be considered as a prosaic act [5] of everyday life of the business and organization contexts and, having into account such diversity, as an empty signifier that cannot be reduced to a single and unequivocal meaning [10].

This dominant discourse of entrepreneurship needs also to be questioned from a gender perspective once it is constituted by a certain normative masculinity that celebrates an individuality [11] inscribed in a heroic and Darwinian model (p. 258) [2] that produces logics of exclusion of those who fail to conform to such masculinity. It is such masculinity centred on individual and intrinsic qualities of the entrepreneurial subject that has been used to neglect its sociocultural meaning and to turn invisible the power gender relations that constitutes it [12, 13]. In this sense, for Ogbor [13] it becomes important to insist on the need to expose the process through which normalization and legitimation of such discourse occur, and the discrimination and devaluation of the others that cannot be defined by such discourse such as women. Women are constituted as the other of the ideal subject of entrepreneurship and this gender fault is also extensive to their enterprises that are perceived as less valued, related with economic sectors of low income like retailing and services, and also accused of not being enough innovator [14, 15].

Such masculine discourse that centres on the individual and emphasises personal attributes, such as competitiveness, autonomy, ambition, risk taking ability and ability to make profit, has gained particular relevance in recent years in the Portuguese context, particularly in the interventions of government and institutional members. In some sense, this discourse aims to promote the economic growth and decrease costs with the Welfare State and social responsibilities, but make individuals responsible for their own employment [16]. The celebration of such discourse by the governments, as some researches indicate [7, 17], also allows to control the meanings that are desirable to associate to the entrepreneurial act and that are controlled by the funding plans made available to certain types of entrepreneurial projects and subjects. So, a government desperate to "shorten" its involvement in the Welfare State [16] provides the right political background to support the initiative of individual entrepreneurs.

This focus on the entrepreneurship initiatives gains visibility in the institutional and statistical national reports [18, 19] and in the European directives to support projects of consulting and training in entrepreneurship. It has also increased the introduction of course units of entrepreneurship in the graduate programs and workshops and seminars with the aim to promote the "entrepreneurial spirit" [20].

In this context, the role played by European institutions in changing the way entrepreneurship is viewed is also worth noting. Under the guidelines and financial support of the European Union, a number of programs have been set up to offer consultancy and training in entrepreneurial matters to women as they are recognized as a disadvantaged social group in such context. As already said, women remain a minority in the business world [21]. In Portugal, women occupy only 4.5 % of the leading positions in the enterprises [1]. Entrepreneurship through self-employment appears also as an effective way to guaranty and simulate the participation of women in economic activity.

Imbued with this entrepreneurial spirit we can better understand the emergence of the training programs in entrepreneurship exclusively for women in the Portuguese context. At the international level, those gender affirmative policies have deserved an interesting and fruitful debate about the positive and negative impact of such policies in creating a true equality of opportunities for women. This kind of policies that have been raised with the liberal feminism allow women to get access to some important individual tools and permission to enter or being accepted in the masculine work and organizational territories. However they do not put in question the masculine discourses that constitute the norm in such contexts and as a consequence, women are still viewed as outsiders of such norm that is not considered problematic and presented as neutral [4, 22, 23].

Having in mind the masculinity norm related with the practice of entrepreneurship and the national context in what concerns the institutional support to promote such practice among women, the present study tried to explore how such normative conception of entrepreneurship affects the institutional agents on how they understand such practice and its subjects and also how they related this with gender affirmative policies. In Portugal, this kind of training programs in entrepreneurship exclusively for women are mostly funded by European Union and in the hand of the Institutional Business Associations.

Considering the information presented above, we invited six of the most important Associations, at national and regional levels, to participate in the study. The chapter proceeds now with the methodology used for the purpose of this exploratory study.

3 Methodology

In the beginning of this study, one of the researchers has contacted the most influential and recognized institutional associations of enterprises and explained our research aims and focus on the participation of women in the entrepreneurship contexts. In these first approaches, the persons contacted felt curious because of our focus of attention on women. Nevertheless, this option was perceived as limiting our own research and its contribution to the understanding of such context. And seemed ultimately to diminish the interest of our interlocutors in the study itself. Such reaction can explain the initial resistance to the study with our interlocutors adopting two main and contradictory explanations to devalue the research: (a) one supported by the argument that equality between men and women was something already conquer and perceived as guaranteed so our study was not really useful and necessary for being somewhat out dated; (b) other supported by the argument that the reduced number of businesswomen did not justified the need for a research project that focused exclusively on their impact and participation on the entrepreneurial and economic national contexts.

Also worth noting was the fact that the study was perceived as understandable because the researcher that contacted the institutions and explained the research aims was also a woman. It seems that it would be easier for the members of those institutions to find a female researcher interested in doing research on "women issues" in entrepreneurship than if the researcher was a man. Such reactions were also more recurrent in members of such organizations that did not presented any initiatives related with the promotion of gender equality or, having such initiatives, were not directly involved.

Six out of 10 institutional associations of enterprises contacted have agreed to participate in this exploratory study. Data was collected through interviews conducted with organizations representing most small and medium enterprises, at national and regional levels: Portuguese Entrepreneurial Association (AEP), Business Association of Minho (AIMINHO), National Association of Businesswomen (ANE), Association of SME-Portugal (PME-Portugal), Commercial Association of Braga (ACB), National Association of Young Entrepreneurs (ANJ). In Portugal, Small and Medium Enterprises are mostly non-financial and for entrepreneurs with individual name, i.e. sole ownerships, are situated in the north region and constitute around 99 % of the national economic tissue [24, 25].

One or two interviews were conducted on each organization with the director or the responsible for the training programs that were being develop and implemented by the associations. In table 1 we present some information about these six associations and the interviewees.

Each participant was interviewed face-to-face following a semi-structured guide in order to explore the system of meanings related to the entrepreneurial subject, their perceived experiences with businesswomen and their entrepreneurial contexts and their positioning in relation to the training programs exclusively for women. The interviews lasted from one to one hour and a half.

The apparent difficulty to deal with the problematic of gender in entrepreneurship became visible in the first institutional contacts and was also present in some interviews. At the beginning of the interviews, during general and introductory questions (related with the general activity of the institutional associations, the categorization of the entrepreneurs and of the national economic and enterprise national tissue) most interviewees gave spontaneous answers and spend time explaining and developing their ideas. When the specific situation of businesswomen and the policies of gender affirmative action on entrepreneurship, like training programs exclusively for women were addressed, they reacted differently. By this time, the interventions seemed more rushed, and sometimes ended in an abrupt way or were deviated for thematic issues apparently neutral to gender and of their own interest. This will become very obvious in the section devoted to data analysis.

In a reverse way, the interviewees that were more involved in gender affirmative policies tended to be more open to the gender issues and to naturalize the research aim. This was easier for them to understand since they made part of the political and institutional framework as professionals responsible for applying gender policies. Nevertheless, such openness did not necessarily meant a great or a better knowledge and reflexivity about gender equal opportunities and discrimination in the entrepreneurship contexts, since their discourses in relation to the gender issues tended to be very similar to the institutional rhetoric transmitted by the European Union.

Table 1 Characterization of the Institutional Associations of Enterprises and the interviewees

Associations	Interviewees	Objectives	Sectors of activity	Organizational size, ownership and region	Female entrepreneurs as business partners (%)	Gender initiatives
ANE	Two interviews: President Former president and founder of ANE	Legal, fiscal and economic information, training and consultancy services	Trade and services	Micro and small entreprises; single-member company (ownership) and private limited companies; North region	100 %	Incentive programs to promote equal opportunities (EQUAL, EXITO and DIVERCIDADE). Specialized training on equal opportunities for both genders (funded courses). Does notpromote training courses in entrepreneurship for women
AEP	Project manager	Lobby; entrepreneurial consultancy and training	Industry (52 %), services (18 %) and trade (23 %).	Small and medium entreprises (65 %); private limited companies; North region	Not able to identify the exact percentage, but the interviewee considered it to be very small	Does not promote specific programs on equal opportunities. By the time the interview was conducted,the association did not promote training courses in entrepreneuship for women
AIMINHO	President	Lobby; entrepreneurial consultancy and training services	Industry (textile, metalworking and construction), services and trade	Small and medium entreprises (95 %); private limited companies; North region	Not able to identify the exact percentage, but the interviewee considered it to be very small	Is about to start a pioneer project in entrepreneurship exclusively for women: Ideas and Business led by Women (funded courses)

(continued)

Table 1 continued

Associations	Interviewees	Objectives	Sectors of activity	Organizational size, ownership and region	Female entrepreneurs as business partners (%)	Gender initiatives
ANJE	Two interviews: Vice-Director; Training coordinator	Lobby; entrepreneurial consultancy and training services	Services and trade	Micro and small entreprises; single-member company (ownership) and private limited companies; North region	A maximum of 20 %	Incentive programs to promote equal opportunities in work-family relationship (funded project). By the time the interview was conducted, the association was starting a training project in entrepreneurship for women: JENE project, a funded course to be held in Lisbon and Algarve
PME-Portugal	Training coordinator	Investment support, consultancy and training services	Construction, services and trade	Micro and small entreprises; single-member company (ownership) and private limited companies; North region	A maximum of 10 %	Training programs in entrepreneurship for women: "Mrs.Entreprise" (funded course)
ACB	Director	Support to trade and to commercial companies; consultancy and training services	Trade	Micro and small entreprises; single-member company (ownership) and private limited companies; North region	About 50 %	Training projects on equal opportunities (funded courses for both men and women). By the time the interview was conducted, the association had just started offering courses in entrepreneurship for women (funded courses)

All the interviews were recorded and transcribed verbatim. Data were organized and coded with the help of a qualitative data management software (Nvivo 8.0). In a first step we tried to identify the system of meanings related to the entrepreneurial act, the ideal entrepreneurial subject, the businesswomen and the training programs exclusively for women in each interview and then identifying the system of meanings shared between all the interviews. In a second moment we intended to explore the tensions and contradictions that emerged between that system of meanings [26] that allowed the interviewees to give coherence to their discourses about entrepreneurship and gender in order to became legitimate representatives in their role of members of an institutional association of entreprise. To accomplish this, we have considered in the discourse analysis of gender in relation to entrepreneurship two main categories: the construction of entrepreneur subject and the positioning towards gender affirmative policies in entrepreneurship.

4 Results

4.1 The "New" Masculine Subject of the Portuguese Entrepreneurship and the Condition of Other of the Businesswoman

In the discourses of our interviewees we can identify a tension between an affirmation of the diversity of the Portuguese entrepreneurs and the presentation of a stable version of the entrepreneur subject. This construction of the entrepreneur subject is marked by the use of time category. A discursive strategy that serves to give an account of the meaning metamorphoses suffered by the entrepreneur subject towards modernity. To achieve that the participants of the study tried to renounce to a discourse of paternalism informed by meanings related with tradition, past, authority and where the business subject is named as "boss". Although the new discourse also centres on the individual, it is constructed around the narrative of modernity that puts emphasis on competence and professionalism, risk, innovation, prosperity and enterprise dimension.

Our chairman likes to oppose the concept of entrepreneur to the concept of the boss (…) I believe that his idea already addresses the need for a bigger dimension, responsibility over the company and the workers. There wasn't in that concept he likes to present, the one of entrepreneur, the notion of the boss, the traditional one, so to speak //I think that the ability to create wealth, to generate more wealth/there are small businesses that, due to their nature and the notion the entrepreneur has of them, will never have that dimension, that importance, that ability to create wealth. //but there is a growing number of those small businesses, I believe there is another type of entrepreneurs/a second generation businessmen, whose parents were the company founders, their children had qualifications, had the opportunity to study and achieve superior qualifications who then began to manage the company (…)

recently there is the notion of a highly qualified student whose initial idea is not to get a permanent employment but to create his own business based in his initiative and all the knowledge acquired during his academic education. (Entrepreneurial Association 2[3])

20 years ago we, 20 years ago we had the businessman who was the boss. It was a different reality, there was not that exposure necessity/in this evolution that took place, the fact of being an entrepreneur or a boss was a fact that had been undermined (…) Let me tell you that in fact, the average level of culture of our entrepreneur was, indeed, low! Now, it is growing exponentially (…) the level of culture of the boss, is still further back, he defended himself because he did not know better, all he wanted was to earn money, his objective was profit, profit, profit///today's entrepreneur is not that person who is obliged to invest in training, he has to do it necessarily/he cannot be with the employee because he likes him or because he is his cousin, no! (Entrepreneurial Association 3)

I have the perception that we are beginning to have a new generation of entrepreneurs with different skills from those 20 or 25 years ago. We are beginning to have better prepared people in terms of their education. With higher education, people who know the world and that is very significant. If the person already acquired certain undeniable business values, such as competition, rivalry, the fact that competition is made at a global level. (Entrepreneurial Association 4)

In this process of giving meaning to entrepreneurship we can see that the interviewees' interventions are centred on the characterization of the individual. We can also witness the construction of an entrepreneur that is based in several logics of exclusion related with age and education, as illustrated by the previous quotes, and also gender. Such logics implied the emergence of several "others". In relation to gender, women and businesswomen are then constructed as the other of such new entrepreneur as a result of their gender condition. Such condition of otherness is not only related with the masculinization of the practice of entrepreneurship (the numeric predominance of men) but also with the masculine symbolic meanings in which such practice is culturally and socially inscribed (the individual traits and values that are perceived as necessary or desirable in the entrepreneur subject, such as flexibility, competence and separation from the private and family contexts).

Entrepreneurial activity is, by definition, an engaging activity, which requires great flexibility and is sometimes incompatible with the inflexibility that family life imposes. This observation is easily sustained from the number of women who assume responsibility of new projects, who participate in training activities oriented for new projects [of entrepreneurship]. (Entrepreneurial Association 4)

Society was created with its difficulties, isn't it? So, today for a woman to be an entrepreneur is easy but 15 or 20 years ago, listen, it was not simple! She had a set, an environment that was not favourable to her, she could not smoke in public, people/she put her thing, her cigarette, isn't it? She is a human being! Equal to

[3] The names were kept confidential to allow a more free use of the quotations and to preserve anonymity.

man! She can also smoke, it is bad for the lungs, spend her money like man does, and so on/but only men could, smoking was exclusive to men. If people decide is basically men, women have no chance "So, what does that chick has? What an odd thing!"/She has a restaurant, she has this/that was the way people though. Nowadays things are starting to change, but the truth is that several years gone by and today the leading positions are clearly dominated by men and entrepreneurs are also men. (Entrepreneurial Association 3)

In the discourses of interviewees the validation of the gender discrimination in the practice of entrepreneurship is not followed by a critical view that questions the power gender relations that give meaning to the contexts of entrepreneurship. In some way, there isn't a reflexivity about the social and symbolic logics that are intrinsic to that masculinity, that result in accepting without critique the demands of "great flexibility" and the externality and strangeness of the "family life" in relation to the entrepreneurship practice. Additionally, those discourses are also characterized by an uncritical exclusive association of women with the family and private contexts. In the second quote there is inclusively a certain normalization of the gender discrimination explained by the history as illustrated in the following passage: "Society was created with its difficulties, isn't it?". Although in an hesitant way and without conviction, it is related with past. Equality is also perceived here as a result of the masculinity of women, and consequently as a means to supress their lack of gender and enabling women to be treated equally (here symbolized in the smoking behaviour, used by the interviewee to illustrate gender equality). This masculinity of women is nevertheless presented in a negative way: by emphasising the loss of health and money and by concluding that if women try to be like men, they will lose the opportunities of being accepted (in the context of entrepreneurship).

A different positioning concerning women in entrepreneurship is to deny their discrimination in such contexts and emphasising its feminization. Nevertheless, in the subtext of such interventions it is not easy to confirm such convictions. The interviewees cannot easily give examples and mention statistics that sustain and corroborate gender equality. A strategy used to overcome the difficulty in substantiating their argument is appealing to personal work experiences that are not totally or directly related with entrepreneurship practice. This is the case of the following interviewee:

For example, maybe this is also a good example. In our Association we have about 30 collaborators, among who are those that are developing their curricular training, their professional training, and we have at the level of the Association, in our organic/in our organization, we have this organized by departments. With the exception/well, the CEO is a man (laughs) of course/but what was I saying with this? Those responsible, for example, for the departments, the Association/financial department, training department and technical assistance of our members, promotion and fairs, are coordinated and assumed by the respective teams, are coordinated and assumed by women. This to say the following: now in companies may be the partners, or the entrepreneur, or the CEO or/may be a man, but the truth is that the intermediate management positions and top management positions are occupied by women. (Entrepreneurial Association 6)

Nevertheless, such quote also informs that the best example that the interviewee choose to confirm the feminization and the increasing gender equality (even if not in entrepreneurship) is actually a very poor one since it only serves to illustrate the feminization in middle management positions. In this example, it is also visible that the top management positions are still being occupied by men. Paradoxically, such example ends up confirming the secondary place of women in the power positions since the chosen examples are not related with entrepreneurship. Such secondary position is paradigmatically explicit in the quotation: "Well, the CEO is a man (laughs) of course".

In general, the persistent idea of increased feminization of the entrepreneurship context is nevertheless followed by a lack of knowledge of the actual participation of women in such context and the interviewees' reference to predominant daily contacts with businessmen and not with businesswomen. We can then conclude that the strategy to emphasise the increase of businesswomen and the tendency for the creation of an equality of gender in entrepreneurship produces a paradoxical effect due to the incapacity of justifying and defending such strategy. The interviewees' inability to substantiate the discourse on equality confirms once again the invisibility of women in such contexts, reinforced by the need to illustrate with individualized references of businesswomen that belong to their professional network.

I have a sporadic knowledge of some cases/Look at the Association's General Management chart and you already know which are the cases of entrepreneur women most present in the life of the Association/Maria Lomba,[4] *is a business partner but is also a business daughter.* (Entrepreneurial Association 2)

Generally speaking, there are an increasingly larger number of women to come forward with business initiatives in the trade area/I'm reminding two specific cases, who are fair dealers, women who perform/who develop business in a City Hall, for example, and are affiliated on this Association over 70 years. Over 70 years, right? I mean, and now their relatives are carrying on that business, isn't it? (Entrepreneurial Association 6)

I know female entrepreneurs from all sectors, so I know a lot of people. Let me tell you that I do not make distinctions. Businesswomen are, at the very least, as fighters as men are (…). Let me also tell you that their academic education is slightly, in mine/from what I know, superior than men's and a woman to become an entrepreneur had to be better prepared, that is my opinion. 20 years ago who was the businesswoman? Ok. Bravo from the airdrome, Fernanda something, then Estrela but the Estrela from the last ten years. Women are more/better prepared, it is a very competitive world in which they have to do more than men do, because women are less considered than men, isn't it? (Entrepreneurial Association 3)

This discursive strategy of making reference to personal cases has a reverse effect, since it emphasises an essentialist version of the businesswomen in which men are still being the normative reference to proceed to the comparison between the two genders in the entrepreneurial context and in relation to the individual qualities/skills that are desirable in this context. Paradoxically, even in presence

[4] All the names are fictitious.

of scarce information about the situation of businesswomen, they are nevertheless described as better than men. Although the most important individual traits in doing business are still ascribed to men, they are more easily assigned to the "profile of the businessman" than to the "profile of a businesswoman"

I have no doubt about one thing/from our experience, oh is that, usually businesswoman is very, is more (pause) how should I say this?! Perhaps more (pause) edgy, isn't it? About the development of her business concept/she is much more (pause) incisive/in trying to achieve certain goals and objectives, isn't she? (…) men in the trade part/it is very important all the perseverance that sometimes is much/much easier even/is much easier to find in a businessman profile than in a businesswoman profile, isn't it? (Entrepreneurial Association 6)

In opposition to that other, emerges the construction of the entrepreneur subject legitimized by a discourse of individuality that turns such figure into the protagonist of the economy and to which is associated a set of individual characteristics and psychological traits, such as high level of education, youth, high self esteem, leadership, risk taking ability and communication skills. The individual profile of the entrepreneur subject also emerges when the interviewees speak about the process of recruitment and selection of the candidates chosen to participate in the training programs exclusively for women. The task to elect who has such attributes or to facilitate the learning process of some of these skills is understood as being under the responsibility of the Institutional Associations of enterprises. The exteriorization of such educational role turns these associations' agents in the legitimation of such masculine discourse of entrepreneurship. And, indirectly, as legitimators of the shortage of women who need to be prepared to "fit" in such discourse of "real" entrepreneurial subjects.

What are we trying to find? Women who came to us with a clear business idea and who show some skills like entrepreneurial and leadership spirit/who have a certain psychological profile, who can withstand the pressure to form a business, that is a very important factor//They are people who believe in their own capabilities, who show leading skills, I believe these are some of the, the personal aspects that, that a businesswoman demonstrates/when there is really, when it shows that they are insecure, without, who will have some difficulty in, in assume a management position, a business management, they can hardly evolve, or evolving a company, I suppose that, that maybe this is it? (Entrepreneurial Association 5)

Ah the entrepreneur is/is mainly someone who takes chances, who assumes the risk of developing an economical activity by/taking into account, oh, the risk, and the risk is/taking risks of/the option rather than being an employee. An entrepreneur is someone who assumes the risk of, of moving towards an entrepreneurial initiative. Regardless of the sector or branch of business, in the perspective of, obviously, improving his or her income, of creating jobs (pause) to proliferate an ideal, isn't it? (Entrepreneurial Association 6)

Our Association has been doing a pioneer work at this level in Portugal and we have also come to the conclusion that being an entrepreneur, of creating a company is something that, somehow, is also cultivated and is done with the help of Portuguese schools. (Entrepreneurial Association 4)

All the concern and interest in the selection and transmission of the individual characteristics and psychological traits with the aim to promote the entrepreneurship in Portugal turned these organizations into active participants in the reproduction of an individualist discursive version of entrepreneurship. And has also contributed to the recognition of the entrepreneur subject gendered by the masculine as one of the important characteristics of the organizational national narratives of modernity. Nevertheless, and paradoxically, the literature about entrepreneurship still makes reference to the relevance of a family culture [27, 28].

The participants' discourse also shows that the system of meanings in which entrepreneurship is inscribed can be intersected by different gender, age, education and generational conditions and by the economic situation of a country. Here, and as remember by Rosslyn Reed [29], the masculinity of entrepreneurship must be understood as dynamic and differentiated in relation to different periods of time and social and economic contexts. We can then distinguish an "old" masculine discourse of entrepreneurship, more traditional, less educated and related to family from a new entrepreneurship, more professional and individualistic.

In the next pages we will further explore the strategies used by the interviewees to constitute their positioning in relation to the training programs of entrepreneurship exclusively for women.

5 The Discursive Positioning in Relation to the Politics of Entrepreneurship for Women

As already said, the entrepreneurship practice has been understood by the governments and by the academy as a political measure to solve the problem of the unemployment and gender discrimination in the workplace [30, 31]. These are also the reasons presented by the participants to describe those who apply to the training programs in entrepreneurship: young women graduated and/or unemployed and women long-term unemployed.

Yes, we have several projects/but is much more tackling unemployment/we contribute to training, but is much more tackling unemployment. (Entrepreneurial Association 1)

Given that the labour market penalises the vast majority of women, so they are the main target of, of unemployment and the majority of unemployed end up being women, the European fund management together with the commissions for equal opportunities thought of this product on, on a first stage this would be a well integrated programme which would allow those unemployment victims, being women, some aged, isn't it? Over 35 years old, had the opportunity to change their professional life through the creation of their own business, and so the programme emerges, which I really believe has many potential. (Entrepreneurial Association 5)

According to Ahl [3], in the Finnish Society, the interest with the role of women on entrepreneurship practice and more especially with policies aimed at promoting their participation in such context, occur mainly by the external influence of European

Union and its directives in gender equality. Such interest also assumes particular instrumental nature, as shown in the discourses of the interviewees. The discursive enunciation assumes two features: (a) in terms of content, by valuing and giving priority to economic and employment factors in detriment of the questions related with the gender discrimination; in terms of form, by the relationship of proximity established by the interviewees between issues related to the access to the fundings of European Union and the candidature to programs of entrepreneurship exclusively for women, usually materialized by training and business consulting.

The candidatures and the development of gender affirmative action in entrepreneurship by the Institutional Associations do not necessarily mean a high reflexivity on gender questions. The discussion about such programs is followed by some constrained and discursive confusion in the articulation of arguments to advocate the link of the Institutional Association to such programs. The following quote illustrates the constraint of one interviewee about training courses that are also exclusively for men (the secretariat courses), which are being implemented by his/her Institutional Association. Such derivation presents this association as an exemption, since it favours women. On the other hand, the questions of sociocultural inequality between genders is less referred and developed.

There are several inputs for this, obviously, isn't that right? On one hand is/are the available resources and should be enhanced. And the Association has, has the role of being the enhancing agent of this/this/those resources, those qualities, and on the other hand, there is/from de people who are unemployed there is/there is a potential, a search/and a search that we/lets say it like this we are receiving, isn't it? We Association, or any institution nowadays (pause)/is the incessant influx of curriculum. And this is an indicator, and we know, so there are evidences/ And encouraging signs that unemployment is starting to (pause)/There are being created, there are being created jobs, but the truth is that we still have a large unemployment rate, we are (pause) in different areas with market integration (...) Is within the support employment and training programme that now ends, isn't it? It prevailed in the framework/in which the Community framework/ah and to exist a proper programme to promote equal opportunities/the access to certain occupations, isn't that true? Within that our Association, like others, have analysed projects, targeting to/to different audiences. I am going to give you a specific example, relative to men because usually secretaries were women. We have proposed to create a training programme to unemployed men in that field of expertise. For them to find an employment opportunity in this area. And it was very successful. We had a large number of men interested in attending that training programme and then/ they have searched that training with success. Also/also within that measure there was a side, a side related to (pause) to training women, specially those who were unemployed. And what was given, what was suggested was a training programme for those women. Ah, that training after which there was a consultancy period, so a specific support for those women, so they could move themselves for a entrepreneurial initiative. (Entrepreneurial Association 6)

Based on the interviews, we can conclude that the measures to promote gender equality produce a paradoxical relation between an argumentation that neglect or disregards the gender inequality issues in favour of economic issues but simultaneously assumes the positioning of agents in promoting and implementing such political gender initiatives. This paradox somehow allows understanding the discourse confusion and constraint felt by the interviewees when the theme of gender affirmative policies in entrepreneurship was approached.

Taken all into account, we could said that assuming an explicit positioning in favour of such policies using a central argumentation based on the questions of inequality in gender relations is not compatible with the implicit acceptance of the masculine norm in which entrepreneurship is inscribed and that is presented as gender-neutral. Assuming a position against such policies is considered impossible because it defies the desirable institutional rhetoric that is promoted by the European Union.

It is also relevant that the interviewees had difficulty in remembering other programs with similar aims that could be implemented by their Institutional Associations, beyond initiatives on gender equality in entrepreneurship that were funded. One of the Associations did however make reference to some measures with the aim to avoid the conflict between family and work responsibilities. Such measures were also exclusively for women since the family is also understood as a major or exclusive responsibility of them. Once again men were not mentioned, being understood as free from such responsibilities.

We have developed a series of projects that we have subjected to financing. Imagine that you want to start your own company. In here we have a space, the facilities, logistic and administrative support and a day care centre. The idea was that women could in a single place conciliate their professional activity with monitoring their children. (Entrepreneurial Association 4)

We can also say that such measures of reformist nature despite creating better conditions for women to get access to territories that are traditionally masculinized and masculine, can involuntarily allow the reproduction of the masculine norms in such contexts. And may further hamper the change of the essentialized version of women as well as the promotion of new ways of (re)thinking the entrepreneurship practice and, in this particular case, the parenting responsibilities and family [2].

6 Discussion and Conclusions

The exploratory study that we have conducted with six institutional associations of enterprises contributes to highlight some central reflections traversing all the discursive records. The first one is the construction of the subject of entrepreneurship inscribed in a masculine discourse of individuality centred on the professionalism, innovation, risk and economic progress that is part of the narrative of modernity. We can say that the construction of such business and organizational subject is dynamic [29] but nevertheless remains masculine.

A second reflection is related with the construction of the businesswomen as the other of such entrepreneur subject. Such otherness is visible in the lack of knowledge about such socioeconomic group and in a tense representation between two positioning: one of the businesswomen as having a "deficiency of gender" [32] for not being able to correspond to the masculine idealization presented as neutral (e.g. when exclusively linked to family; or when men are perceived as more perseverant and assertive in the commercial area); the other by women masculinization in order to justify their presence in such contexts. In both cases we can witness the essentialization of businesswomen that remains trapped by the sociocultural gender constraints.

So, the condition of otherness of the businesswomen constrained their recognition as legitimate and proper subjects in the context of entrepreneurship. That also helps to explain why this new attention given to businesswomen, as a result of such training programs of entrepreneurship exclusively for women, are somehow presented as a "novelty" and a "discovery" that denies implicitly their active presence in such context through all these years and that is very previous to this institutional interest. Often, as referred by Bruni et al. [2] the women who participate actively in the life of their organizations, sometimes with their husbands are denied the sociocultural ideal model that distinguish the entrepreneur subjects, and are only defined through their family relationships (as wives of the businessman) [27].

A third reflection is related to the gender reformist policies, more specifically to the gender affirmative actions in entrepreneurship taken by the institution associations of enterprises. In this case we can said that there is not a clear interiorization of such policies by these institutions. The motivation is essentially external and is often the result of an "outside" imposition of the European Union. The adoption of such policies has also an instrumental nature because it centres firstly on the access to European Union financial funding and on the economic and unemployment effects of such policies. They are not intended to give priority and change the gender power relations in the entrepreneurship contexts. Such aims are turned secondary or ignored in the discourses of the interviewees. So, this exploratory study contributes also to call attention to a paradox between the purposes of gender affirmative actions related to entrepreneurship and the lower reflexivity in the gender issues of the institutional agents that implemented such policies. It is then important to rethink the role of such polices in the promotion of a real gender equality. As written elsewhere [4], if such policies can contribute to force the acceptance and entrance of women in such context and provide them some individual tools to compete with men in entrepreneurship, it questions their condition of otherness and their marginal and secondary positioning. To do that is important to put the focus and also to deconstruct the logics of exclusion that are legitimated even if involuntarily by those who are interested in promoting the economic participation and the employment between women through the entrepreneurship. Maybe it would be important to change not women but the conceptions of gender and entrepreneurship of the institutional agents and also men that belong and assume an active role in these contexts. In this order of ideas, maybe it would be important to create training programs to these different players of the entrepreneurial context to change their construction of the gender relations.

References

1. INE. (2012). *As empresas em Portugal 2010*. Lisboa: Instituto Nacional de Estatística.
2. Bruni, A., Gherardi, S., & Poggio, B. (2004). Entrepreneur-mentality, gender, and the study of women entrepreneurs. *Journal of Organizational Change Management, 17*(3), 256–268.
3. Ahl, H. (2002). The construction of the female entrepreneur as the other. In H. Hopfl & B. Czarniawska (Eds.), *Casting the other, the production and maintenance of inequalities in work organizations* (pp. 52–67). New York: Routledge.
4. Ahl, H. (2004). *The scientific reproduction of gender inequality: A discourse analysis of research texts on women's entrepreneurship*. Malmo, Sweden: Copenhagen Business School Press.
5. Steyaert, C. (2004). The prosaics of entrepreneurship. In D. Hjorth & C. Steyaert (Eds.), *Narrative and discursive approaches in entrepreneurship: A second movements in entrepreneurship book* (pp. 8–21). Cheltenham, UK: Edward Elgar.
6. Steyaert, C., & Katz, J. A. (2004). Reclaiming the space of entrepreneurship in society: Geographical, discursive and social dimensions. *Entrepreneurship & Regional Development, 6*(3), 179–196.
7. da Costa, A., & Saraiva, L. (2012). Hegemonic discourses on entrepreneurship as an ideological mechanism for the reproduction of capital. *Organization, 19*(5), 587–614.
8. Essers, C., Doorewaard, H., & Benschop, Y. (2013). Family ties: Migrant female business owners doing identity work on the public-private divide. *Human Relations, 66*(12), 1645–1665.
9. Gill, R. (2014). If you're struggling to survive day-to-day: Class optimism and contradiction in entrepreneurship discourse. *Organization, 21*(1), 50–67.
10. Jones, C., & Spicer, A. (2005). The sublime object of entrepreneurship. *Organization, 12*(2), 223–246.
11. Ahl, H., & Marlow, S. (2012). Exploring the dynamics of gender, feminism and entrepreneurship: Advancing debate to escape a dead end? *Organization, 19*(5), 543–562.
12. Harvey, A. M. (2005). Becoming entrepreneurs. Intersections of race, class, and gender at the black beauty salon. *Gender & Society, 19*(6), 789–808.
13. Ogbor, J. (2000). Mythicizing and reification in entrepreneurial discourse: Ideology-critique of entrepreneurial studies. *Journal of Management Studies, 37*(5), 605–635.
14. Anna, A. L., Chandler, G. N., & Mero, N. P. (1999). Women business owners in traditional and non-traditional industrial. *Journal of Business Venturing, 15*, 275–303.
15. Mitra, R. (2002). The growth pattern of women-run enterprises: An empirical study in India. *Journal of Development Entrepreneurship, 7*(2), 217–237.
16. Peren, L., & Jennings, P. L. (2005). Government discourses on entrepreneurship: Issues of legitimisation, subjugation, and power. *Entrepreneurship Theory and Practice, 29*(2), 173–184.
17. Kenny, K., & Scriver, S. (2012). Dangerously empty? Hegemony and the construction of the Irish entrepreneur. *Organization, 19*(5), 615–633.
18. CITE. (2009). *Relatório sobre o progresso da igualdade de oportunidades entre mulheres e homens no trabalho, no emprego e na formação profissional—2006–2008*. Lisboa: Comissão para a Igualdade no Trabalho e no Emprego.
19. INE (2009).O empreendedorismo em Portugal. Indicadores sobre a demografia das empresas. Lisboa: Instituto Nacional de Estatística.
20. Redford, D. (2006). Entrepreneurship education in Portugal: 2004/2005 national survey. *Comportamento Organizacional e Gestão, 12*(1), 19–41.
21. Amway Europe (2013). Amway global entrepreneurship report 2013.
22. Bruni, A., Gherardi, S., & Poggio, B. (2005). *Gender and entrepreneurship, an ethnographical approach*. New York: Routledge.
23. Calás, M., Smircich, L., & Bourne, K. (2009). Extending the boundaries: Reframing 'entrepreneurship as social change' through feminist perspectives. *Academy of Management Review, 34*(3), 552–569.
24. INE. (2012). *Ser mulher em Portugal, 2001–2011*. Lisboa: Instituto Nacional de Estatística.
25. INE. (2013). *As empresas em Portugal 2011*. Lisboa: Instituto Nacional de Estatística.

26. Kilduff, M., & Kelemen, M. (2004). Deconstructing discourse. In D. Grant, C. Hardy, C. Oswick, & L. Putman (Eds.), *Organizational discourse* (pp. 259–298). London: Sage Publications.
27. Guerreiro, M. das D. (1996). Famílias na actividade empresarial. *PME em Portugal*. Oeiras: Celta Editora.
28. Lima, M. A. P. d. (2003). *Grandes famílias, grandes empresas: ensaio antropológico sobre uma elite de Lisboa*. Lisboa: Publicações Dom Quixote.
29. Reed, R. (1996). Entrepreneurialism and paternalism in Australian management: A gender critique of the 'self-made' man. In D. Collinson & J. Hearn (Eds.), *Men as managers, managers as men: Critical perspectives on men, masculinities and managements* (pp. 99–122). London: Sage Publications.
30. Hughes, K., D. (2003). Pushed or pulled? Women's entry into self-employment and small business ownership. *Gender, Work and Organization*, 10(4), 433–454.
31. Mattis, M. C. (2004). Women entrepreneurs: Out from under the glass ceiling. *Women in Management Review, 19*(3), 154–163.
32. Fournier, V. (2002). Keeping the veil of otherness, practising disconnection. In H. Hopfl & B. Czarniawska (Eds.), *Casting the other, the production and maintenance of inequalities in work organizations* (pp. 68–85). London: Routledge.

Influence of Organizational Flexibility in High Performance Work Practices

Mercedes Rubio-Andrés and Santiago Gutiérrez-Broncano

Abstract High-performance work practices (HPWP) can help to create an organizational effectiveness and therefore a sustainable competitive advantage in firms. Literature shows considerable evidence that these practices are associated with firm performance. Besides, organizational flexibility may moderate the relationship between HPWP and firm performance, and also has a more positive effect on firm performance. Our study proposes a research model which analyzes the relationships between performance practices, organizational flexibility and firm performance. We also try to study the determinants of organizational flexibility as the key to compete in turbulent and dynamic environments.

1 Introduction

Research focusing on the firm-level impact of HRM practices has become popular in recent years (for reviews, see [1–5]). In general, Researchers have reported positive associations between measures of firm-level systems of human resource management and organizational performance [3, 4, 6–9].

In addition, the literature includes studies that focus on the performance effects of specific HRM practices, such as training [10, 11] and information sharing [12, 13], and research that examines the influence of systems of such practices on organizational outcomes [3, 4, 8, 9]. Although many studies have reported a

M. Rubio-Andrés (✉)
Department of Business, School of Business and Economics, San Pablo CEU University, C/Julián Romea 23, 1.08 B, 28003 Madrid, Spain
e-mail: mrubio@ceu.es

S. Gutiérrez-Broncano
Faculty of Social Sciences, Department of Business Administration, University of Castilla-La Mancha, Avda. Real Fábrica de Seda, s/n, 45600 Talavera de la Reina, Toledo, Spain

C. Machado and J. P. Davim (eds.), *Work Organization and Human Resource Management*, Management and Industrial Engineering,
DOI: 10.1007/978-3-319-06376-8_8,

positive association between various HRM practices and objective and perceptual measures of firm performance, some authors [5, 14] have expressed concern that results may be biased because of methodological problems. In addition, the absence of a widely accepted measure of the "high performance" HRM practices construct makes it difficult to compare findings across studies (for examples of different approaches, see [1, 3, 4, 7, 9].

A conceptual work has also argued that complementarities, or synergies, both among a firm's HRM practices and between a firm's HRM practices and its competitive strategy, can have an additional and positive effect on firm performance [15, 16].

2 Objectives

- To analyze the importance high performance practices and a positive association between various HRM practices and firm performance
- To study the determinants of organizational flexibility
- To check the association with high performance practices, organizational flexibility and firm performance.

3 Determinants of Organizational Flexibility

Flexibility is the ability of a firm to respond to various demands from its dynamic competitive environment [17].

But what determines whether or not an organization is flexible?

It is possible to talk about flexibility in various concepts: numerical flexibility [18]. Functional flexibility [19, 20], flexible workforce [21], flexible manufacturing systems [9] and organizational flexibility [22].

Research into a firm's organizational flexibility suggests at least two possible sets of determinants of flexibility: first, those determinants concerning a flexible structural design [23]; and second, those regarding new managerial capabilities required in a flexible firm [24].

Flexible structural design means attaining a structure that allows the flexible organization tosucceed under environmental pressure and unpredictability [25–27]. Krijnen [28] and Overholt [29] point out that a flexible structure requires decentralization in decision-making, low levels of formalization, a high degree of permeability of boundaries, collaborative partnerships, delayering of business units and autonomy [30].

The second set of determinants indicated in the literature on organizational flexibility are those concerning the new managerial capabilities needed to succeed in fast-changing environments. Volberda [24] assert the importance of a broad knowledge base and a broad scope of managerial expertise in order to devise appropriate responses. The heterogeneity in backgrounds and experiences needed in a flexible firm is related to the need to face competitive environments. More hetero geneous managerial expertise may enhance the absorptive capacity of the organization for recognizing the need for change [24, 31].

Finally, Sanchez [17] notes that there are two basic types of flexibility: resource flexibility and coordination flexibility. Resource flexibility refers to the extent to which a resource can be applied to a larger range of alternative uses, the costs and difficulty of switching the use of are source from one alternative use to another, and the time required to switch from one use to another.

Coordination flexibility consists of the extent to which the firm can resynthesize the strategy, reconfigure the chain of resources, and redeploy the resources. These types of flexibility are particularly applicable for exploring the concept of flexibility in strategic HRM. Wrigth and Snell [32] apply concepts of resource and coordination flexibility to HRM practices, employee skills, and employee behaviors. In essence, we can broadly conceive of flexibility in strategic HRM as the extent to which the firm's human resources possess skills and behavioral repertoires that can give a firm options ior pursuing strategic alternatives in the firm's competitive environment, as well as the extent to which the necessary HRM practices can be identified, developed, and implemented quickly to maximize the flexibilities inherent in those human resources.

4 High Performance Work Practices: Dimensions

High performance practices [33–35] or high involvement practices [3], are also known by some authors as 'soft' managements models of human resources [36, 37]. These terms essentially refer to the planning of human resources management oriented towards the commitment of the employees, involving the active participation of these in decision making and providing the necessary organization support and resources [38]. It stems from the idea that the effect produced between capabilities and motivation has a multiplying effect on the value created in the organization [39].

There are two principal advantages for organizations in implementing systems of human resources management oriented towards performance: firstly, it enables the organization to be more effective [33] and secondly, it increases the social acceptance of this model in comparison with the traditional taylorist style based on strict control and subordination of the employees [40].

Researchers have attempted to determine which best human resources practices have generated the best results by grouping non-traditional practices that have been increasingly used in companies. Ichniowski et al. [41] cite work flexibility, teamwork, contingent remuneration, empowerment, job security, etc., as practices that generate the greatest level of commitment among employees.

Marchington and Grugulis [42] presented a model based on the seven practices identified by Pfeffer [43] as generating success: job security, selective contracting of personnel, teamwork, contingent compensation, extensive training, reduction of status differences and transfer of information.

Table 1 Determinants of organizational flexibility

Flexible structure—"loose structure"	• Descentralization in decision-making • Low levels of formalization • High degree of permeability of boundaries and collaborative partnerships • Delayering of business units • High levels of individual and team level • Autonomy
New managerial capabilities	• Heterogeneity • Broad managerial mindsets
Employee flexibility	• Behavior flexiblitity • Skill flexibility

Source authors

Implementing all of these human resources practices has a profound impact on the employees and the teams because it increases his/her potential, motivation and commitment to the company [17], all of which are fundamental elements for improving innovation.

The commitment generated within the company makes the employee act in a positive manner, which translates into an effective way of achieving organizational objectives and leads to improved business results [49, 51, 56]. HR Practice flexibility are significantly associated with an index of firm performance Besides, Literature in HPWP shows considerable evidence that these practices are associated with organizational performance [1, 3, 6, 9, 10, 55] because they render higher levels of motivation, satisfaction, commitment, and production in the employees and promote a more effective firm.

5 Effects of Organizational Flexibility Between High Performance Practice and Performance

Many scholars agree that moderators and mediators exist between HRM practices and firm performance. In a meta-analysis, findings by Combs et al. [59] suggest that organizational strategy and context may moderate the relationship between HRM and firm performance (Tables 1, 2, 3).

Hage [60], Hage and Aiken [61, 62], Perrow [63], Lawrence and Lorsch [64], all of them pointing to the arousal of flexible structures with fuzzy roles and horizontal communication channels (hence, HR practices) whenever the environment is sufficiently uncertain and unpredictable.

In one empirical study, Youndt et al. [46] find that organizational manufacturing strategy, such as cost strategy, quality strategy, and flexibility strategy, moderates the HR practices and firm performance relationship. Combs et al. [59] also suggest that employees' knowledge, skills, and abilities (KSAs) and the social structure act as mediators between high performance work practices and organizational performance. Guest [36] shows that strategic HRM (SHRM) improves firm performance by enhancing employees' skills and abilities.

Table 2 Principal high performance human resources practices

High performance practices	Description	Principal authors
Creating positions with wide scope	Jobs with a wide scope, low horizontal and vertical specializations	Schuler and Jackson [44], Capelli and Crocker-Hefter [45]
Selective recruitment	Defined recruitment and selection criteria appropriate to the demands of the position. Analysis of past, present and future conduct	Guest [36], Arthur [6], Pfeffer [43], Huselid [3], Youndt et al. [46], Roche [47], Bayo and Merino [48], Ordiz and Fenández [49], Céspedes et al. [50], Beltrán et al. [51], Wood and Menezes [52]
Fixed contracting	Job stability to avoid temporary work	Pfeffer [43], Bayo and Merino [48], Ordiz and Fernández [49], Wood and Menezes [52]
Extensive training	Great importance and investment in the training and development of the employee. Continuous and evaluated training of all employees. Incentivise skill acquisition	Schuler and Jackson [44]
Performance evaluation	Evaluation schemes based on results	Guest [36], Pfeffer [43], Huselid [3], Huselid et al. [8], Capelli and Neumark [53], Bayo and Merino [48], Richard and Johnson [54], Guthrie et al. [55], Roca et al. [38], Mohr and Zoghi [56], Beltrán et al. [51], Wood and Menezes [52]
Contingent compensation	Salary policies based on individual and group results. Reward above the market average. Tangible, intangible and flexible incentives	Guest [36], Arthur [6], Pfeffer [43], Huselid [3], Roche [47], Bayo and Merino [48], Capelli and Neumark [53], Guthrie et al. [55], Ordiz and Fernández [49], Céspedes et al. [50], Zatzick e Iverson [57], Beltrán et al. [51], Wood and Menezes [52]
Greater discretion	Award the position and all its responsibilities to the employee in order to evaluate conduct and norms in the job. Decentralization and less vertical specialization	Schuler and Jackson [44, 58], Arthur [6], Pfeffer [43], Huselid [3], Roche [47], Bayo and Merino [48], Richard and Johnson [54], Guthrie et al. [55], Ordiz and Fernández [49], Roca et al. [38], Mohr and Zoghi [56]
Internal promotion	Provide opportunities for professional development in the company through a formal system of professional careers	Arthur [6], Pfeffer [43], Huselid [3], Roche [47], Bayo and Merino [48], Richard and Johnson [54], Guthrie et al. [55], Roca et al. [38], Ordiz and Fernández [49], Mohr and Yoghi [56]

(continued)

Table 2 continued

High performance practices	Description	Principal authors
Teamwork	Incentivize and promote teamwork, employee commitment and Little job rotation	Arthur [6], Pfeffer [43], Huselid et al. [8], Bayo and Merino [48], Capelli and Neumark [53], Richard and Johnson [54], Guthrie et al. [55], Zatzick e Iverson [57], Mohr and Zoghi [56]
Shared information	Open and shared information systems for all members of the organization	Guest [36], Arthur [6], Pfeffer [43], Huselid et al. [8], Roche [47], Capelli and Neumark [53], Bayo and Merino [48], Richard and Johnson [54], Guthrie et al. [55], Ordiz and Fernández [49], Roca et al. [38], Zatzick e Iverson [57], Mohr andZoghi [56], Wood and Menezes [52]
Job security	High levels of job security. Specialized staff to improve job security	Schuler and Jackson [44, 58]

Source authors, based on Martí [69]

Table 3 High performance practice, organizational flexibility and performance

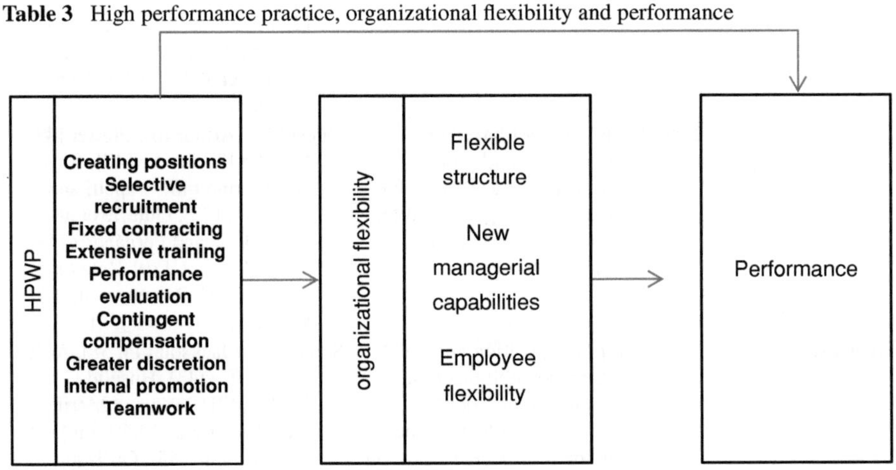

Source authors

Other researchers have shown that flexibility in other functional areas of the firm, such as operational flexibility, product customization, and resource flexibility is related to increased firm performance [9, 65–68], HR flexibility and its possible contribution to firm performance and competitive advantage.

Our proposal is that further organizational flexibility, has a moderating role and also has a more positive effect on firm performance.

6 Conclusions

Literature suggests that HR flexibility is a dynamic capability facilitating a firm's rapid response to changing environments. High performance practices are positively related to firm performance. Flexible structure or "loose structure", new managerial capabilities and Employee flexibility has a moderating role and also has a more effect on firm performance.

References

1. Appelbaum, E., & Batt, R. (1994). *The new American workplace: Transforming work systems in the United States*. Ithaca: ILR Press.
2. Berg, P., Appelbaum, E., Bailey, T., and Kalleberg, A. (1994). The performance effects of modular production in the apparel industry. *Working Paper*. Washington, DC: Economic Policy Institute.
3. Huselid, M. A. (1995). The impact of human resource management practices on turnover, productivity, and corporate financial performance. *Academy of Management Journal, 38*, 635–670.
4. Ichniowski, C., Shaw, K., and Prennushi, G. (1994). The effects of human resource management practices on productivity. *Working Paper*. New York: Columbia University.
5. Wagner, J. A. (1994). Participation's effect on performance and satisfaction: A reconsideration of research evidence. *Academy of Management Review, 19*, 312–330.
6. Arthur, J. B. (1994). Effects of human resource systems on manufacturing performance and turn—over. *Academy of Management Journal, 37*, 670–687.
7. Cutcher-Gershenfeld, J. (1991). The impact on economic performance of a transformation in industrial relations. *Industrial and Labor Relations Review, 44*, 241–260.
8. Huselid, M. A., & Becker, B. E. (1994). The strategic impact of human resources: Results from a panel study. *Working Paper*. New Brunswick, NJ: Rutgers University.
9. MacDuffie, J. P. (1995). Human resource bundles and manufacturing performance: Flexible production systems in the world auto industry. *Industrial and Labor Relations Review, 48*, 197–221.
10. Bartel, A. P. (1994). Productivity gains from the implementation of employee training programs. *Industrial Relations, 33*, 411–425.
11. Knoke, D., & Kalleberg, A. L. (1994). Job training in U.S. organizations. *American Sociological Review, 59*, 537–546.
12. Kleiner, M. M., & Bouillon, M. L. (1988). Providing business information to production workers: Correlates of compensation and profitability. *Industrial and Labor Relations Review, 41*, 605–617.
13. Morishima, M. (1991). Information sharing and firm performance in Japan. *Industrial Relations, 30*, 37–61.
14. Levine, D., & Tyson, L. D. (1990). Participation, productivity, and the firm's environment. In A. S. Blinder (Ed.), *Paying for Productivity* (pp. 183–244). Washington: Brookings Institution.
15. Baird, L., & Meshoulam, I. (1988). Managing two fits of strategic human resource management. *Academy of Management Review, 13*, 116–128.
16. Milgrom, P., & Roberts, J. (1995). Complementarities and fit: Strategy, structure, and organizational change in manufacturing. *Journal of Accounting and Economics, 19*, 179–208.
17. Sanchez, R. (1995). Strategic flexibility in product competition. *Strategic Management Journal, 16*, 135–159.
18. Atkinson, J., & Meager, N. (1986). Is flexibility just a flash in the pan? *Personnel Management, 18*(9), 26.

19. Smith, V. (1994). New forms of work organization. *Annual Review of Sociology, 23*, 315–329.
20. Hunter, L., McGregor, A., MacInnes, J., & Sproull, A. (1993). The "flexible firm": Strategy and segmentation. *British Journal of Industrial Relations, 31*, 387–407.
21. Dastmalchian, A. (2001). Workplace flexibility and the changing nature of work: An introduction. *Revue Canadienne de Sciences del'Administration, 18*, 1–4.
22. Teece, D. J., Pisano, G., & Shuen, A. (1997). Dynamic capabilities and strategic management. *Strategic Management Journal, 18*, 509–533.
23. Englehardt, C. S., & Simmons, P. R. (2002). Organizational flexibility for a changing world. *Leadership and Organization Development Journal, 23*, 113–122.
24. Volberda, H. W. (1999). *Building the Flexible Firm*. Oxford: Oxford University Press.
25. Ackoff, R. L. (1977). Towards Flexible Organizations: A multidimensional design. *OMEGA, 5*, 649–662.
26. Eppink, J. (1978). Planning for strategic flexibility. *Long Range Planning, 11*, 9–15.
27. Foss, N. J. (2003). Selective intervention and internal hybrids: Interpreting and learning from the rise and decline of the Oticon Spaghetti Organization. *Organization Science, 14*, 331–349.
28. Krijnen, H. C. (1979). The flexible firm. *Long Range Planning, 12*, 63–75.
29. Overholt, M. H. (1997). Flexible Organizations: Using organizational design as a competitive advantage. *Human Resources Planning, 20*, 22–32.
30. Palmer, I., Benveniste, J., & Dunford, R. (2007). New organizational forms: Towards a generative dialogue. *Organization Studies, 28*(12), 1829–1847.
31. Calori, R., Baden-Fuller, C., & Hunt, B. (2000). Managing change at Novotel: Back to the future. *Long Range Planning, 33*, 779–804.
32. Wrigth, P. M., & Snell, S. A. (1998). Toward a unifying framework for exploring fit and flexibility in strategic human resource management. *Academy of Management Review, 23*(4), 756–772.
33. Lawler, E. E. (1992).The ultimate advantage: Creating the High-Involvement Organization. San Francisco: Jossey-Bass.
34. Walton, R. E. (1985). From control to commitment in the workplace. *Harvard Business Review, 63*, 77–84.
35. Wood, S. Albanase, M. T. (1995). Can we speak of a high commitment management on the shop floor?. *Journal of Management Studies, 32*, 215–247.
36. Guest, D. E. (1987). Human resource management and industrial relations. *Journal of Management Studies, 24*(5), 503–521.
37. Truss, C., Gratton, L., Hope-Hailey, V., McGovern, P., & Stiles, P. (1997). Soft and hard models of human management: A reappraisal. *Journal of Management Studies, 34*, 53–73.
38. Roca Puig, V., Escrig Tena, A. B., & Bou Llusar, J. C. (2002). Compromiso con los empleados y estrategia competitiva: Un análisis intersectorial de su repercusión en los resultados" *Cuadernos de Economía y Dirección de la Empresa, 12*, 267–289.
39. Ulrich, D. (1998). Intellectual capital = competence × commitment. *Sloan Management Journal, 4*, 15–26.
40. Sashkin, M. (1984). Participative management is an ethical imperative. *Organizational Dynamics, 12*(4), 5–22.
41. Ichniowski, C., Kochan, T. A., Levine, D., Olson, C., & Strauss, G. (1996). What works at work: Overview and assessment. *Industrial Relations, 35*, 332–352.
42. Marchington, M., & Grugulis, I. (2000). Best practice human resource management: Perfect opportunity or dangerous illusion? *The International Journal of Human Resource Management, 11*(6), 1104–1121.
43. Pfeffer, J. (1994). Competitive advantage through people. *California Management Review, 36*(2), 9–28.
44. Schuler, R. S., and Jackson, S. E. (1987a). Linking competitive strategy with human resource management practices. *Academy of Management Executive, 1*(3), 207–219.

45. Cappelli, P., and Crocker-Hefter, A. (1996). Distinctive human resources are firm score competencies. *Organizational Dynamics, 24*(3), 7–22.
46. Youndt, M. A., Snell, S. A., Dean, J. W, Jr, & Lepak, D. P. (1996). Human resource management, manufacturing strategy, and firm performance. *Academy of Management Journal, 39*, 836–866.
47. Roche, W. K. (1999). In search of commitment-oriented human resource management practices and the conditions that sustain them. *Journal of Management Studies, 36*, 653–678.
48. Bayo, A., & Merino, J. (2001). Quality management and high performance work practices: Do they coexist? *International Journal of Production Economics, 73*(3), 251–260.
49. Ordiz Fuertes, M., & Fernández Sánchez, E. (2003). High-involvement practices in human resource management: Concept and factors that motive their adoption. *International Journal of Human Resource Management, 14*(4), 511–529.
50. Céspedes, J. J., Jerez, P., & Valle, R. (2005). Las prácticas de Recursos Humanos de Alto Rendimiento y la capacidad de aprendizaje organizativo: Incidencia e implicaciones. Cuadernos de Economía y Dirección de la Empresa, nº 24, pp.29–56.
51. Beltrán M., I., Roca P., V., Escrig T., A., & Bou L., J., C. (2008). Human resource flexibility as a mediating variable between high performance work systems and performance. *Journal of Management, 34*(5), 1009–1044.
52. Wood, S., and de Menezes, L. M. (2008). Comparing perspectives on high involvement management and organizational performance across the British economy. *The International Journal of Human Resource Management, 19*(4), 639–682.
53. Capelli, P., & Neumark, D. (2001). Do high-performance work practices improve establishment level outcomes?. *Industrial and Labour Relations Review, 54*(4), 737–775.
54. Richard, O. C., and Johnson, N. B. (2001). Strategic human resource management effectiveness and firm performance. *The International Journal of Human resource Management*, 12, 299–310.
55. Guthrie, J. P., Spell, C. S., & Nyamori, R. O. (2002). Correlates and consequences of high involvement work practices: The role of competitive strategy. *International Journal of Human resource Management, 13*(1), 183–197.
56. Mohr, R. D., & Zoghi, C. (2008). The high-involvement work design and job satisfaction. *Industrial and Labor Relations Review, 61*(3), 275–296.
57. Zatzick, C. D., & Iverson, R. D. (2006). High-involvement management and work force reduction: Competitive advantage or disadvantage? *Academy of Management Journal, 49*(5), 999–1015.
58. Schuler, R. S., and Jackson, S. E. (1987b). Organizational strategy and organization level as determinants of human resource management practices. *Human Resource Planning, 10*(3), 125–141.
59. Combs, J., Liu, Y., Hall, A., & Ketchen, D. (2006). How much do high-performance work practices matter? Ameta-analysis of their effects on organizational performance. *Personnel Psychology, 59*, 501–528.
60. Hage, J. (1965). An axiomatic theory of organizations. *Administrative Science Quarterly, 10*(3), 289–320.
61. Hage, J., & Aiken, M. (1969). Routine technology, social structure, and organization goals. *Administrative Science Quarterly, 14*(3), 366–376.
62. Hage, J., & Aiken, M. (1970). *Social change in complex organizations*. New York: Random House.
63. Perrow, C. (1967). A framework for the comparative analysis of organizations. *American Sociological Review, 32*(2), 194–208.
64. Lawrence, P. R., & Lorsch, J. W. (1967). *Organization and Environment*. Boston: Harvard University, Graduate School of Business Administration.
65. Garud, R., and Kotha, S. (1994). Using the brain as a metaphor to model flexible production systems. *Academy of Management Review, 19*, 671–698.

66. Parthasarthy, R., and Sethi, S. P. (1993). Relating strategy and structure to flexible automation: A test of fit and performance implications. *Strategic Management Journal*, *14*, 529–549.
67. Rangan, S. (1998). Do multinationals operate flexibly? Theory and evidence. *Journal of International Business Studies, 29*, 217–237.
68. Thomke, S. (1998). Agile product development: Managing development flexibility in uncertain environments. *California Management Review, 41*(1), 8–30.
69. Martí, N. (2008). High Commitment Human Resource Practices in Spanish Call Centres. Nature, Sources and Effects on Voluntary Turnover. Thesis Doctoral Universitat Rovirai Virgili.

Gaining or Losing? Projective Identification, Professional Identities and New Public Management

Teresa Carla Trigo de Oliveira and João Fontes da Costa

Abstract Professional identities can be challenged by radical organisational change. This chapter suggests that this is especially the case when the change is ideologically driven such as New Public Management (NPM) in health reforms. It submits that concern to achieve greater efficiency through economies of scale and mergers between hospitals and services has displaced limits to these in the sphere of production. The chapter relates this to Kleinian projective identification of market values and displacement and denial of realties in NPM reforms. It also shows from in depth discourse and its analysis with health professionals in two major European hospitals whose services have been merged by NPM reforms that recognising differences in professional identities and allowing relative autonomy for them at operational levels is needed for both economic efficiency and also social efficiency in terms of the wellbeing of health professionals and of patients.

1 Introduction

Major organizational change not only can challenge professional identities and roles but may intend to do so. As was the case with New Public Management (NPM) in the UK [1–3] which drew on claims that health professionals needed recognise the case for 'quasi-markets' and private sector performance criteria to achieve greater efficiency in its National Health Service. Yet also may prove

T. C. T. de Oliveira (✉)
Faculty of Economics, University of Coimbra, Av. Dias da Silva 165,
3004-512 Coimbra, Portugal
e-mail: tcarla@fe.uc.pt

J. F. da Costa
Estgoh, Polytechnic Institute of Coimbra, Rua General Santos Costa,
3400-124 Oliveira do Hospital, Portugal

C. Machado and J. P. Davim (eds.), *Work Organization and Human Resource Management*, Management and Industrial Engineering,
DOI: 10.1007/978-3-319-06376-8_9,

dysfunctional in terms of both economic and social efficiency where the former concerns cost reduction and the latter concerns both the wellbeing of the public and of health professionals in delivering an efficient health service. Whereas, a more recent approach such as that of Osborne, Radnor, and Nasi [4] known as New Public Service Management (NPSM) has been trying to redress this. While others such as Brown and Yates [5] have submitted that effective reforms of public services should combine an overall design with more empowerment and more autonomous decision-making at operational levels.

This chapter first addresses this in relation to the difference between institutional, organisational and operational logics in health reforms and the limits of drawing on paradigms of work organization from the sphere of production which were stressed by Oliveira and Holland [6] and recently have been paralleled and extended by Osborne Radnor and Nasi [4]. Second, it submits that the drive for market criteria and values may be ideologically driven in a manner that projectively identifies people with capital, displaces the role of knowledge, experience and skills, and may deny that NPM style reforms are not delivering their intended results. Third, it relates this also to issues concerning professional identities and the degree to which treating health professionals as factors of production fails to recognise the diversity of such knowledge, abilities and skills and the need for NPSM style reforms to achieve a degree of relative autonomy at operational levels for them to be able to deploy them effectively. Fourth, it cites challenge to professional identities by NPM reforms in a southern European country, which has projectively identified with the UK NPM case while less than consciously displacing that this can be dysfunctional in terms of both economic and social efficiency. It then draws some findings, discussion of the findings implications for further research into New Public Service Management (NPSM).

2 Institutions, Organizational Change and Logics

Throughout Western Europe there is increasing demand for health services with ageing populations at the same time as governments are seeking to reduce the costs of health provision by the introduction of 'quasi-markets' [1–3] on the presumption that these also will enhance both patient choice and social equity [7].

However, within the explicit logic of seeking to achieve greater economic efficiency are implicit logics that rarely are surfaced in debates on NPM. Such that seeking quasi-markets on the presumption of gains from economies of scale—more *of* the same—displaces economies of scope—more *with* the same. This is paralleled in the sphere of production by the difference between Ford's mass production and Taylor's ultra division of labour and post Fordism in the sense of the Toyota production model [8–10].

The Ford-Taylor synthesis initially gained spectacular success. Within a few years Ford was producing more than half of the world's motor vehicles and, in the interim, eliminated more than 300 craft producers of them in the US. Yet its management model was top-down, and inflexible. Ford could not delegate, nor was prepared to consider improvements to his Model T, or the case that it should have any colour other than black. A few surviving producers in the US then were regrouped in General Motors where an accountant, Alfred Sloan, created a multi-divisional management model [9, 11, 12].

Yet which then succumbed to an inertial organisational logic in assuming that this combined with Ford's economies of scale and Taylor's ultra-division of labour was the height of organisational efficiency and, like Taylor's 'one best way', could not be improved. Whereas it was not. A small Japanese company which was on the brink of failure after WW2 came to overtake GM as the world's no 1 vehicle producer by combining economies of scope and flexible, skill based production with what it later achieved in terms of economies of scale [8].

Within which there are parallels with what NPM has been aiming to achieve in health services, which has been based on presuming to achieve Fordist economies of scale while neglecting the degree to which there both are differences between economies of scale and of scope, and between the sphere of production in the private sector, and the aims of a social service such as health provision in the public domain [4, 13]. These also can be expressed in terms of different institutional, organisational and operational logics.

Thus a national health service is an institution in which a government lays down norms, protocols and rules concerning professional roles, patient care and economic performance. The institution is pyramidal ranging from finance ministries as well as health ministries, through regional health authorities and health areas to local practice. A hospital is an organisation closer to the multi-divisional model by which Sloan at General Motors overcame Ford's own refusal to delegate. Just as large companies are organised in production, finance, marketing, sales and other divisions, so hospitals are organised in departments. Each tends to have its own operating theatre, its own diagnostic unit, its own research, its own wards and its own nursing and other technical medical staff [10].

However, while Sloan endorsed Taylor's [14] task specialisation for managers and line workers, health professionals in a major and complex organisations such as a general or teaching hospital are multi-tasked and need to be multi-skilled. With only a few exceptions, which may include anaesthetics, specialisation in hospitals tends to be intra-departmental. Gynaecology and cardiac surgery may share some techniques and technologies, but are not identical. Their specialists need to be near the frontier of knowledge in their area. In a good teaching hospital, they will be on or advancing the frontier of best practice.

Such specialisation and the range of different tasks and skills within a hospital suggests the case for relative autonomy of decision-making similar to that of a holding company rather than the hierarchical top-down single model design and performance assessment presumed by New Public Management [15]. Such a hierarchical design underlay the introduction of NHS reforms in the UK, with increased concern with economic performance in terms of reducing unit costs, bringing in managers with private sector experience at regional and area levels and introducing new layers of supervisory management to monitor performance on the grounds that health professionals were self-interested and could not be trusted [16].

There also is a presumption in some other countries that the New Public Management reforms in the UK NHS (and, since devolution, in England) offer a template for change management. There is less awareness outside the UK that these have revealed conflicting rationalities between government insistence on more market criteria in health provision and opposition to this from health professionals, or that criticism of the reforms for lack of consultation on how they have been introduced has been persistently voiced by the British Medical Association, the Royal College of Nurses; the Royal College of General Practitioners and the Royal College of Midwives [6, 13, 17, 18].

The explicit logic of the reforms had been to devolve responsibility and increase efficiency by privatising or out-sourcing services. But their implicit logic has been Weberian in terms of pyramidal top-down authority, Fordist in its preoccupation to increase patient throughput to reduce unit costs, and Taylorist in terms of constant surveillance of performance criteria and outcomes [19]. While the outcome, with increased layers of supervisory management, has not reduced but trebled administrative costs as a share of total costs from under 5 % to over 14 % [18].

More recently, Osborne, Radnor, and Nasi [4] have argued that NPM is not fit for purpose, submitting that it contains two fatal flaws. First, that it focuses on intra-organizational processes when the reality of public services delivery is interrelational. Second, that it draws upon management theory derived from the production experience of the manufacturing sector and which ignores differences between this and public and social services. They submit that health care, education, community development and regeneration, for example, are not concerned with final product delivery such as of a car or a household good, but are relational, interactive and ongoing between people, and therefore need different institutional and organisational logics.

3 Market Values and Social Values

Different institutional logics in New Public Management (NPM) also relate to different perceptions of market values and social values. These are stylised in Fig. 1. The figure distinguishes explicit market values from implicit social values.

On its left, the closer one 'gets to market' the more explicit is an exchange relationship, as with a contract to buy or sell, or of employment. The more one is concerned with property rights, the more explicit are the terms and conditions of such rights, again in formal contracts. The value of human and social rights also may be explicit, as in codes of conduct against ethnic or gender discrimination, or a charter of universal rights, such as the UN Charter of Human Rights and the right to wellbeing in the sense of the UNDP criteria of health, longevity, education and standard of living. But it is characteristic of social values that they tend to be tacit and implicit in norms of behaviour rather than rules, protocols or codes, nor reducible to a single value, such as economic efficiency rather than social welfare and wellbeing.

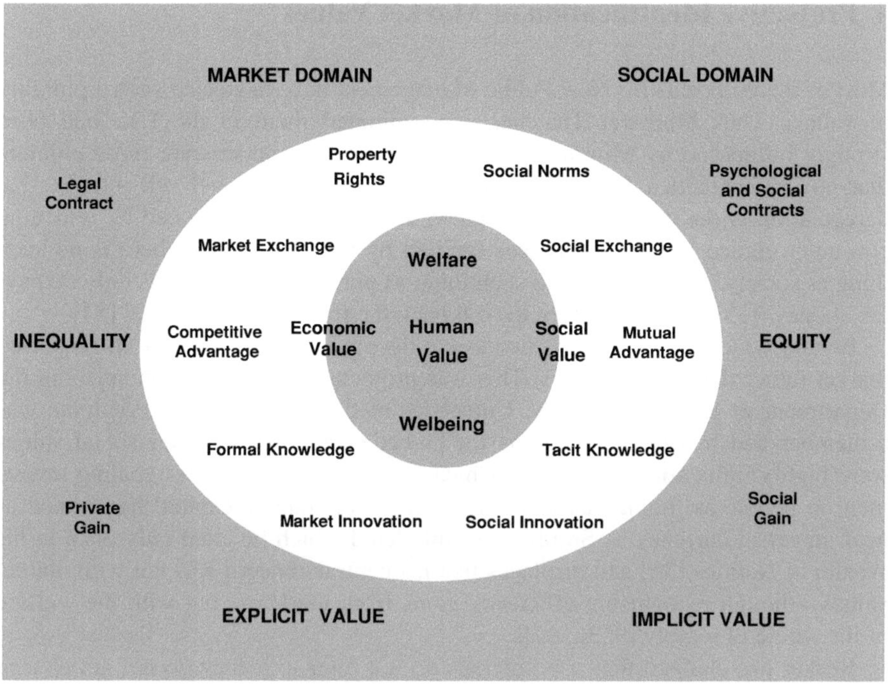

Fig. 1 Market domains and social domains [9]

Sen's approach to welfare and values parallels this. What he proposed instead is the need to recognise a plurality of values [20] and making way for 'a variety of human acts and states as important in themselves' [21, p. 33]. This includes not only hedonism in the sense of pleasure rather than only self-gratification but the Aristotelian principle of *eudaimonia* or purposeful living, which is consistent with Rousseau's stress on overcoming individual, social and political alienation [22] and what Riesman [23] has distinguished as self-direction rather than other-direction. What we are suggesting in line with this is the following:

- Market values sanction inequality, and tend to promote it. Human value relates centrally to issues of equity and of procedural and distributive justice, as in the manner stressed by Sen [24].
- Shared social values enable functional societies and may enhance capabilities [24], whereas lack of them may prove dysfunctional.
- Social value has its own implicit logic in terms of tacit rules and implicit norms of behaviour [13] rather than is explicit in the manner of market exchange.
- There is social value for functional psychological and social contracts and dysfunction from breach of them as submitted by Denise Rousseau and others [25–29].
- Institutions which have social use value, such as that public health services, may be of more valued both by the public and for their providers when they are not centrally market based or profit driven.

4 Projective Identification of Market Values

Most of the proposals for New Public Management have neglected such a plurality of values. Thus Margaret Thatcher, who initiated them in the UK, had been strongly influenced by Milton Friedman claiming that markets are more efficient than governments; that they will maximise social welfare through self-interest; that deregulation is the basis of a free society, and that this is enhanced by widening consumer choice [30–32]. This was typified by her claims that 'there is no such thing as society', that 'there is no such thing as public money; there is only taxpayers' money' [33] and her concern to 'roll back the frontiers of the State' [34].

Behind which, in turn, there also was a deeply ideological approach favouring market rather than social values. This was projected by several economists in the Department of Economics of the University of Chicago of which Friedman was a member and in which his claims for precedence of market over social values were highly influential. It included Chicago's Gary Becker re-designating investment in people as 'human capital'. As a metaphor 'human capital' has gained as near universal currency as Smith's invisible hand which he cited only once in his Wealth of Nations [35] and displaces that his central concern was not with market values—though recognising efficiency gains from markets—but with the welfare of the whole of society [9, 36, 37].

Becker has claimed that if politicians do not refer to it they do not get elected [38]. They may be less aware that he also claimed that slavery was the only market that accurately priced it (idem). Nor that in order to correlate such 'human capital' investment in education with later employee income, Becker needed a range of in human assumptions such as that 'all workers are identical' (ibid. p. 301). Without this he could not attribute differences in earning trajectories entirely to the different investment costs of secondary and higher education. Whereas assuming that all workers are identical denies the differences in personality, skill or experience which individuate people [19].

Becker [38] declined to value or measure experience despite its importance for employers, who rank it highly both in selection and employment since he found 'too difficult to measure' and therefore did not try. He disregarded any education or training that neither is formally structured, nor involves financial investment. He also presumed diminishing returns from investment in 'human capital' (ibid. p. 312) despite the degree to which learning from work and life experience may increase over time [13].

Becker further neglected the degree to which different methods of work organisation may facilitate or frustrate operational or organisational learning. He implicitly assumed non-learning organisations as his organisational paradigm, in which it is formal education, R&D, technology and a high division of labour determines output per worker. As he puts it: '[t] he growth in capital depends on investments in new technologies, basic research and human capital' [39, p. 311]. Yet by excluding experience and informal on-the-job training, he thereby made no allowance for factors which may be centrally important to informal skill acquisition, insight and innovation in a 'knowledge economy' or a 'learning organisation'.

Presuming to advance on Becker, theorists of intellectual capital such as Edvinsson and Sullivan [40] and Stewart [41] presume a range of inputs to what they deem to be 'intellectual capital' including investment costs in formal education, as with Becker, and also 'outputs' such as patents and licences registered by an organisation. They then formalised this as follows in claiming that Intellectual Capital (IC) = Human Capital (HC) + Structural Intellectual Capital (SIC) + Relational Intellectual Capital (RIC).

But, deconstructing the abstractions of these presumed 'Capitals', reveals that the advocates of these concepts consistently identify them as follows:

IC Knowledge, abilities, skills and experience
HC Individual knowledge, abilities, skills and experience
SIC Knowledge, abilities, skills and experience within an organisation
RIC Knowledge, abilities, skills and experience between organisations.

Granted this consistent claim for knowledge, abilities, skills and experience, there is no analytic, methodological or taxonomic justification to designate them as capital. These are human attainments and attributes, and those who deploy them in organisations are labour, nor capital [6].

Yet within this there was another agenda. For instance, Edvinsson and Sullivan [40] and Stewart [41, 42] have claimed that intellectual capital is 'knowledge which can be converted into value' where the value they intended is not for society but for shareholders. Which is consistent with the ideological shift since the 1980s from a presumption that a national health service should be for society and based on social values rather than market values.

Moreover, as Arnold and Cohen [43] have observed, while metaphor has potential to provoke new understandings, 'some metaphors become so diffuse that they cease to be valuable as heuristic devices' [43, ibidem, p. 11]. A metaphor such as 'human capital' can be a simulacrum in the sense of 'a thing having the appearance but not the substance or proper qualities of something; a deceptive imitation or fetishism as in an 'image or representation of a thing, especially a god' (Oxford Shorter Dictionary).

Metaphors also, in terms of Melanie Klein's [44–46] conceptual framework, may combine displacement and denial with idealised—but ungrounded—projective identification. For Klein this meant splitting what is perceived to be good from what is perceived or sensed to be bad, and projecting either onto someone or something else. She developed this in the context of child psychology yet this later was seen by several psychologists to be relevant to splitting from and denial of any unwelcome realities.

Thus Schneider [47] presciently related splitting and projective identification to behaviour on markets, of which an example is rating agencies splitting from recognition that subprime and other financial derivatives could be toxic and projecting them as safe as government bonds. Dinnerstein [48] has extended Kleinian splitting in terms of 'splits between heart and head, feeling and reason, private and public and where 'private is deemed good and public bad' which is precisely what Friedman and the Chicago school have projected.

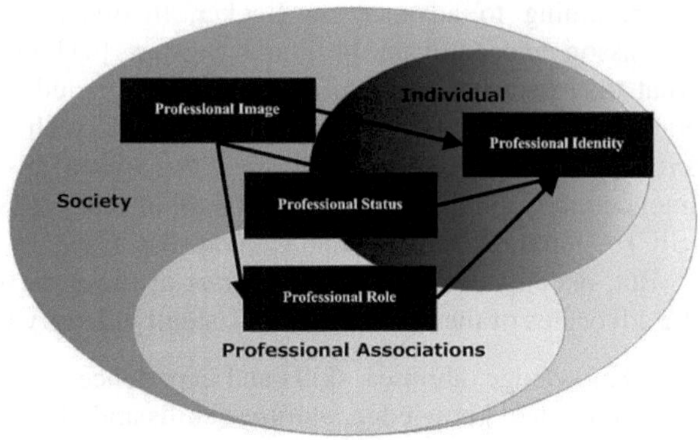

Fig. 2 Society, Professional Associations and professional identity, own elaboration

5 Professional Identities and Psychological and Social Contracts

As modelled in Fig. 2, an individual's professional identity, whether as a doctor, nurse, teacher, lawyer or accountant is influenced by interrelated factors including not only professional training, professional status and professional roles but also societal expectations. Yet, as with social values (Fig. 1) such identity cannot be captured only by explicit criteria. It also tends to depend on tacit rules and implicit norms of professional behaviour [13].

Further, in terms of 'enacted identities' Van den Berg, Richardson and Eastman [49] have argued that institutional norms and practices are not operationally meaningful until the individual perceives them as something important to her or his 'well-being'. Contextual influences on individual choices are explicit or implicit in people's narratives [50, pp. 234]. In such narratives individuals and groups may use discourse that symmetrises, at varying levels of consciousness [51] with the values and norms of their work environment [52–54].

Bowen and Ostroff [55], Nishii and Wright [56] and Nishii, Lepak and Schneider [57] have stressed that *current* perceptions relate to *prior* personal dispositions and values—i.e. who we have become from earlier life and work experience—while Grant and Hofmann [58] recognise that interpersonal dynamics influence social expectations and role redefinition.

Bourdieu [59] has distinguished dispositional, situational, practical and normative logics in his concept of 'habitus'. Thus a dispositional logic concerns what we are disposed to think or want, a situational logic where we are and the constraints this may impose on when we either speak out or choose to remain silent, while a practical logic is how things are done, and a normative logic how it is presumed that they ought to be done. Yet none of which may be explicit rather than implicit not least in group behaviour [60].

Functional groups also rely on implicit trust, which relates to concepts of psychological contract. Schein [61] conceptualised this as the how people are committed to an organisation depends on their implicit expectations of what it will give them, and what they owe to it, and whether this is matched by its expectations of what it will give in return. Denise Rousseau [25, 26, 62] with others [e.g. 28, 29], has focused on perceptions of breach of psychological contract by an individual employee rather than a group. Yet, consistent with Bartlett's [60] stress that the group is the basis of knowledge formation, such breach can be perceived jointly at group levels, such as professional groups, especially if what is being demanded of them by higher management, or by a government, comes to be perceived as undermining their personal and professional values.

Inversely, Denise Rousseau's more renowned namesake, Jean-Jacques Rousseau [22] focused on the conditions by which a group or small society could achieve positive outcomes from a social contract. He wanted this to be written, so that there could be a term of reference for those who later might come to question it. But he also recognised that the dialogue that could promote it would identify and surface what otherwise were tacit values, stressing dialogue as a means of surfacing what people knew they valued but had not made explicit [22]. This was long before the more recent theory of surfacing tacit knowledge in management theory [63, 64]. John Locke also allowed that a social contract could involve tacitly allowing the exercise of power by others unless they abused it [65]. Which arguably has been the case in the imposition, without dialogue, of market criteria and market values into national health services in a manner that health professionals perceive to be displacement and denial of their personal values and professional identities.

6 The Case Study

6.1 Research Context

The case study was with doctors and nurses in three hospitals in one urban area which the government in 2011 had insisted should merge on presumption of thereby achieving economies of scale. The largest (teaching) hospital would integrate what formerly were 42 different management areas, services and units into 7. A medium sized general hospital would need in due course to close its emergency unit and related services. A third paediatric hospital would gain new larger premises but also be integrated in the new administrative framework rather than retain its former autonomy. This was within the national context that the government, in May 2011, introduced a range of new criteria to monitor the performance of health professionals much on the lines of NPM in health and hospital reforms in the UK. The three merged hospitals, which served the central region of the country, would integrate some 8,000 health professionals and administrators (1,800 doctors; 2,700 nurses; 40 administrators and 3,460 technical, administrative and auxiliary staff).

6.2 Data Collection

The collection on a socio-cognitive basis was informed by grounded theory principles [66]. As a method developed among others by Charmaz [67, 68], and Henwood and Pidgeon [69], it involves observation, as well as discourse analysis. While the interviews were not a spoken questionnaire they involved a series of opening questions relating to the introduction of NPM. The interviews addressed four main questions: (1) What are the main changes that have taken place in the organization over the past two years that have affected how you do your work? (2) To what do you attribute these changes? (3) What has been the impact of these changes on your ability to provide a high quality of service to patients? (4) How have the changes affected you personally and your well-being at work?

6.3 Data Analysis

Each interview, of from 45 top 90 min, with 18 senior doctors, most of whom were heads of services and units, and with 16 chief nurses, was fully audio-recorded. Interview transcripts were analysed on a template basis for qualitative analysis and the evolution of data driven concepts [67–70].

7 Findings

- *Top-down reorganisation*

Understandably a senior administrator was enthusiastic about the reorganisation and the manner in which this enabled fortnightly meetings of the heads of only seven services or management areas.

> These Management Areas meet every other week with the Council of Administration (Management Board) in a manner which enables an ongoing overview of the hospital. This is very positive.
> Now, for example, we are able to contract with the management areas, in meetings with all the service directors. With this, we are beginning to see what really can be done at the first, second and third levels. Not only in terms of personalising services, but including them such a way we can integrate better.
>
> Director of the teaching hospital

It also was recognised by medical directors that top-down overview was needed not only of one hospital but of several hospitals in a region.

> Demography is variable and there is a desertification in some areas. If there is not an overview of what different hospitals are doing, no one knows well what is being done where. There is duplication between adjacent hospitals … while in others there is out-contracting of services which could be internalised.
>
> Directorof a unit within the teaching hospital

There also were different perceptions of alleged gains from mergers of hospitals and economies of scale.

> We are smaller than the teaching hospital and they are going to close our emergency service. But it is more accessible both because it is closer to local people and, because it is smaller, I could deal with an emergency in minutes, whereas the teaching hospital has an emergency staff of 100 people, but with a triage system that means that if you are a non-critical emergency you may have to wait for hours to be assessed.
>
> Former Clinical Director of the General Hospital

> Instead of treating people here and now we are shuttling them between the three hospitals in ambulances taking time and raising costs that in many cases are not needed.
>
> Former Clinical Director of the General Hospital

> There are cultural differences that relate also to the quality of the service. For instance, we not only are smaller than the teaching hospital but our whole approach has been more multi-disciplinary, which has gains for the quality of diagnosis and treatment.
>
> Director of Audio-Visual Services of the General Hospital

> The merger of different hospitals implies that there should be can integration of specializations. But on some days children are operated on in the teaching hospital and on others in the paediatric hospital… It's all very disorganized and we no longer know who is managing whom.
>
> Nurse in the Paediatric Hospital

> The main change is in professional roles. It used to be the case that anyone in paediatrics specialised in dealing with children. He or she was a paediatrician. Now I have specialists from the main hospital who have different training and experience. I need to make them aware that dealing with children is different and this is difficult.
>
> Senior Doctor in the Paediatric Hospital

- *Reinforced rather than reduced bureaucracy*

It also was recognised by some medical directors that there was a case for the merger of the three hospitals in the local area:

> It makes sense to merge when you have a hospital near to another one of which is short of beds and the other has spare capacity.
>
> Director of a main unit within the teaching hospital

But medical directors also were emphatic that the reorganisation of a hospital into fewer units did not redress whether such reorganisation still was bureaucratic to a degree which inhibited either economic or social efficiency. As one medical director put it:

> The bureaucracy and time wasting claimed to be needed for cost control is unbelievable. If I need even a simple item like a cover for an examination table I can need fourteen different signatures for approval even before the order is made.

Also that:

> Some elementary principles are ignored, such as bulk purchasing or not realising that a drug company is simply re-pricing and rebranding a product rather than offering something that has clinical value.

And that:

> Central control of costs can raise them because the people concerned have no idea of what is needed for clinical safety. For example, I am obliged to order new surgical instruments when practice has shown that their sterilisation can mean their re-use without patient risk.
>
> Doctor in the cardiac unit of the teaching hospital

Which also was echoed by nurses:

> If we want to solve even a relatively simple management problem we have to wait for an immense amount of time while its implications are worked out for all three hospitals. Previously, one just picked up a phone.
>
> Nurse in the Paediatric Hospital

> In the new framework it not only is very difficult to get responses quickly… In many cases we don't know who is managing us.
>
> Nurse in the Paediatric Hospital

Medical directors also deplored that centralised decision-making was reducing autonomy in decision-making.

> What is needed is delegation of competences and responsibilities in areas such as contracting personnel, and to give rewards and incentives.
>
> Senior Doctor in the radiology unit of the teaching hospital

- *Technical progress, costs and social efficiency*

Senior doctors also stressed that no proper estimates were made of technical progress in equipment in relation to the quality of service to patients.

> Some technical progress is revolutionary… But in other cases the costs of equipment may double but their clinical gains are marginal - as little as 5 per cent - and when such an increase is not needed for quality diagnosis or quality patient care and means less money for other more basic purchases.
>
> Doctor in the cardiac unit of the teaching hospital

- *Concern with throughput inhibiting social efficiency in terms of patient well-being*

The director of a service put the case, consistent with Weberian hierarchy and For distacceleration of throughput, that preoccupation with economic efficiency could inhibit patient well-being:

> All the thinking still is top-down and dominated by how many patients we treat. But the outcome is dysfunctional. A report is written in radiology that a woman needs a mastectomy. She then is passed down as on a production line to surgery without ever meeting the surgeon who will perform the operation. No time is allowed for counselling her because this would raise costs. Yet her psychological well-being is vital to her chances of a good recovery.
>
> Doctor in the radiology unit of the teaching hospital

- *Loss of wellbeing for health professionals*

In one of the interviews, the medical director of a main service within the teaching hospital volunteered that a central problem for it was absenteeism.

> A 'total burn out' with people who then neither have the patience for nor are disposed to do anything more.
>
> Doctor in the radiology unit of the teaching hospital

The same medical director also volunteered that more criteria for performance appraisal did not address such problems. What did was motivation and group identity.

> All evaluations are subjective…. What interests me is not only whether a job is well done and on time. I am interested in the degree to which they are committed to excellent performance by the service.

And, in a critique of the limits of top-down performance appraisal consistent with the thesis that there are limits to a production model as a paradigm for health [4]:

> People are not machines. People are much more complex than machines. It is different in managing a factory where the machines are supposed to produce 50 vases by the end of the day and, if they do not do so, someone has to be responsible for that. A hospital is different. It is dealing with people…'

- *Commitment and psychological contract*

The same medical director, when being introduced to the concept of psychological contract, responded immediately that this was what was missing in performance evaluation, was needed, and what he was trying to achieve in relation to his own unit and team working:

> What is absolutely fundamental is to give people positive and serious encouragement in terms of terms of their mission, of which incentives are part. To give them responsibility both to produce and to communicate… to speak with them, to show them…

- *Confusion of Professional Identities*

Whereas Brown andYates [5] have submitted that effective reforms of public services should mean empowerment and more autonomous personalized decision-making, NPM reforms have meant disempowerment and less autonomy in decision-making with a loss of both professional roles and identities.

> The main change is professional roles and experience. It used to be the case that anyone in paediatrics specialised in dealing with children. He or she was a paediatrician. Now I have specialists from the main hospital who have different training and experience.I need to make them aware that dealing with children is different and this is difficult.
>
> Director of a Paediatric Unit

> Before the merger we were more autonomous. Now both we as professionals and patients are losing\ out.
>
> Director of Audio-Visual Services of the General Hospital

- *Situational logic and identities*

> Teams functioned pretty well in the other smaller location because we were closer together. But now, the new bigger operational environment is creating difficulties both in organization and relating to each other.
>
> Nurse in the Paediatric Hospital

> The physical environment and scale of the new building is alienating. Being spread out constrains operational efficiency, such as that on daily basis we used to meet in certain areas both to socialize and to discuss problems and do not do so any more.
>
> Nurse in the Paediatric Hospital

- *Lack of Recognition*

We don't want to be treated differently but for it to be recognised that, as a paediatric hospital, we are different from other hospitals.

Nurse in the Paediatric Hospital

We would like to think that there is a perceptions within the new framework that our values, our performance and out personal investment is recognised… but it is not.

Nurse in the new Paediatric Hospital

There also was concern that pressures to reduce costs was compromising the social efficiency of the service and denying equity which had been one of the main principles of introducing a national health service in the first place.

> Of course we have to pay attention to costs. But a national health service is not a market. It has to be concerned with equity, and not only with whether or not people can pay.
>
> Senior Doctor in the Emergency Unit of the General Hospital

8 Implications of the Findings

Many of the implications of the findings from the case studies have been indicated by grouping them within sub-headings which reflect the conceptual frameworks outlined earlier in this chapter. Many also confirm earlier critiques of the scope and limits of NPM in the British and then English NHS such as by Pollock [17] and Leys and Player [18] as well as those by Oliveira and Holland [6, 19] and more recently by Osborne et al. [4].

They also tend to confirm that ideologically driven NPM reforms have displaced not only that these may not increase economic efficiency but may reduce social efficiency both in the sense of wellbeing for the public and compromise the social and equity values to which health professionals were committed as doctors, or nurses. They challenge the presumption of Le Grand [16] and others that they cannot be trusted and need to be disciplined, by performance related criteria which in practice, in the British case, have raised administrative costs rather than reduced them, and also have profoundly alienated health professionals

They further suggest confirmation that: (1) a Kleinian projective identification with market values displaces the degree to which health professionals are deeply committed to social values and that these are integral to their sense of professional identity; (2) the presumption of intellectual capital theory that employees are a resource to be exploited rather than their knowledge, skills and experience recognised is counter-productive; (3) a Fordist assumption that merging hospitals and health areas will assure cost reductions from economies of scale not only disregards different operational needs and realities but may promote diseconomies; (4) the logic of demanding rigid adherence to Taylorist performance criteria is perceived as lack of trust in health professionals which can be perceived as breach of psychological and social contract and may prove dysfunctional in terms of economic and social efficiency for them and for patients.

9 Implications for Future Research

Where there are implications for further research these are that this should be informed on a grounded theory basis by what actually has occurred in terms of introducing market criteria and market values into social services such as health provision. They could extend the case of Osborne et al. [4] by further consideration of what they submit is the relational nature of a service rather than the final delivery of a product and their case for New Public Service Management (NPSM). As well as further assess the case of Brown and Yates [5] that effective reforms of public services should combine an overall design with more empowerment and greater relative autonomy for decision-making at operational levels.

But also seek deeper analysis of the ideological drive for market based NPM in the claim of Milton Friedman that the private sector always is more efficient than public provision. Since there is next to no evidence that Friedman supported this by any analysis of the relative merits of public and private services rather than mere assertion, and that the same was the case with the Thatcher endorsement of market values rather than social values in initiating what proved to be dysfunctional NPM in the British NHS.

References

1. Bartlett, W., and Le Grand, J. (1993). The theory of quasi-markets, in quasi-markets and social policy Basingstoke. In J. Le Grand and W. Bartlett (Eds.), Basingstoke: Macmillan, pp. 13–34.
2. Propper, C. (1993). Quasi-markets, contracts and quality in health and social care. In J. Le Grand and W. Bartlett (Eds.), *Quasi-markets and social policy* (pp. 35–67). Basingstoke: Macmillan.
3. Bartlett, W., and L. Harrison (1993). Quasi-markets and the national health service reforms. In J. Le Grand and W. Bartlett (Eds.), *Quasi-markets and social policy* (pp. 68–92). Basingstoke: Macmillan.
4. Osborne, S. P., Radnor, Z., & Nasi, G. (2013). A new theory for public service management? Toward a (public) service-dominant approach. *The American Review of Public Administration, 43*, 135–158.
5. Brown, T., and Yates, N. (2010). Choice, empowerment, personalisation: Taking forward public services in a recession. In J. Richardson (Ed.), *From recession to renewal: The impact of the financial crisis on public services and local government* (pp. 157–178). Bristol: Policy Press.

6. Oliveira, T. C., & Holland, S. (2007). Alternative paradigms of Hospital Work Organisation and health provision. *Revista Portuguesa de Saúde Publica, 1*, 19–37.
7. Le Grand, J. (1991). *Equity and choice: An essay in economics and applied philosophy*. London: HarperCollins.
8. Womack, J. P., Jones, D. T., & Roos, D. (1990). *The machine that changed the world: The story of lean production—Toyota's secret weapon in the global car wars that is now revolutionizing world industry*. New York: Free Press.
9. Oliveira, T. C., & Holland, S. (2012). On the centrality of human value. *Journal of Economic Methodology, 19*(2), 221–141.
10. Oliveira, T. C., and Holland, S. (2013). Institutional logics promoting and inhibiting innovation. In C. Machado and J. P. Davim (Eds.), *Management and engineering innovation* (pp. 179–216). IGI Global Associates: Hersey.
11. Lacey, R. (1987). *Ford: The men and the machine*. London: Heinemann and Pan Books.
12. Sloan, A. (1964). *My years with general motors*. New York: Doubleday.
13. Oliveira, T. C. (2007). Delving down to learn up. In *Plenary Presentation to the Sloan MIT-Portugal Conference on New Developments in Management, Lisbon*.
14. Taylor, F. W. (1911). *The Principles of Scientific Management*. Manhattan: Harper & Brothers.
15. Oliveira, T. C., et al. (2013). Hierarchies and holdings: Implications of SME entrepreneurship for enhanced HRM in hospital management. In C. Machado and P. Melo (Eds.), *Effective human resources management in small and medium enterprises: Global perspectives*, Hersey: IGI Global.
16. Le Grand, J. (1997). Knights, knaves or pawns? Human behaviour and social policy. *The Journal of Social Policy, 26*(2), 149–169.
17. Pollock, A. M. (2004). *NHS plc*. London: Verso Books.
18. Leys, C., & Player, S. (2011). *The plot against the NHS*. London: Merlin.
19. Oliveira, T. C., & Holland, S. (2007). Alternative paradigms of hospital work organisation and health provision. *Revista Portuguesa de Saúde Pública, 1*, 19–37.
20. Sen, A. (1987). *On ethics and economics*. Oxford: Basil Blackwell.
21. Sen, A. (1993) Capability and well-being. In M. C. Nussbaum and A. Sen (Eds.), *The quality of life* (pp. 30–53). Oxford University Press and The United Nations University.
22. Rousseau, J. -J. (1960). Du contrat social. 1762—republished in 1960, Paris: Garnier.
23. Riesman, D. (1954). *Individualism reconsidered*. Glencoe: The Free Press.
24. Sen, A. (2009). *The idea of justice*. London: Allen Lane.
25. Rousseau, D. M. (1995). Psychological Contracts in Organizations: Understanding Written and Unwritten Agreements. Thousand Oaks: Sage, Publications Inc.
26. Rousseau, D. M. (1998). Why workers still identify with organizations. *Journal of Organizational Behavior, 19*(3), 217–233.
27. Rousseau, D. M. (1998). The "problem" of the psychological contract considered. *Journal of Organizational Behaviour, 19*(1), 665–671.
28. Rousseau, D. M., and Parks, J. M. (1993). The contracts of individuals and organizations. In L. L. Cummings and B. W. Staw (Eds.), *Research in organizational behaviour*. Greenwich: JAI Press.
29. Rousseau, D. M., and Tinsley, C. (1997) Human resources are local: Society and social contracts in a global economy. In N. Anderson and P. Herriot (Eds.), *International handbook of selection and assessment* (pp. 123–145). New York: Wiley.
30. Friedman, M., & Friedman, R. (1980). *Free to choose*. New York: Harcourt Brace Jovanovitch.
31. Friedman, M. (1962). *Capitalism and freedom*. Chicago: University of Chicago Press.
32. Friedman, M. (1970). The social responsibility of the firm is to increase its profits. In The New York Times magazine.
33. Thatcher, M. (1983). Speech to the Conservative Party Conference.
34. Seddon, J., and O'Donovan, B. (2010) Efficiencies in public service delivery. In J. Richardson (Ed.), *From recession to renewal: The impact of the financial crisis on public services and local government* (pp. 135–156). Bristol: Policy Press.

35. Smith, A. (1910). An enquiry into the nature and causes of the wealth of nations. 1776, republished 1910. London: Dent.
36. Smith, A. (1979). The theory of moral sentiments. In R. D. D. and A. L. Macfie (Eds.), *The theory of moral sentiments,* 1759, reedition 1979. Oxford: Oxford University Press.
37. Rothschild, E. (2000). *Economic sentiments: Adam Smith, Condorcet and the enlightenment.* Cambridge MA: Harvard University Press.
38. Becker, G. (1964). *Human capital.* Chicago: University of Chicago Press.
39. Becker, G. (1993). Human capital. In *A theoretical and empirical analysis with special emphasis on education.* Chicago: The University of Chicago Press.
40. Edvinsson, L., & Sullivan, P. (1996). Developing a model for managing intellectual capital. *European Management Journal, 14*(4), 356–364.
41. Stewart, T. A. (1999). *Intellectual capital: The New Wealth of Organisations.* New York: Doubleday.
42. Stewart, T. A. (1997). Brainpower: Who owns it… How they profit from it. In *Fortune magazine* (pp. 104–110).
43. Arnold, C., & Cohen, L. (2010). Have careers become boundary less? *Human Relations, 63,* 1157–1175.
44. Klein, M. (1932). *The psycho-analysis of children.* London: Hogarth.
45. Klein, M. (1952). *Developments in psychoanalysis.* London: Hogarth.
46. Klein, M. (1984). Narrative of a child analysis. In R. Money-Kyrle (Ed.), *The writings of Melanie Klein.* New York: Free Press.
47. Schneider, M. (1975). *Neurosis and civilization.* New York: Seabury Press.
48. Dinnerstein, D. (1978). *The rocking of the cradle and the ruling of the world.* London: Souvenir Press.
49. Vandenberg, R. J., Richardson, H. A., & Eastman, L. J. (1999). The impact of high involvement work processes on organizational effectiveness: A second-order latent variable. *Group and Organizational Management, 24,* 300–339.
50. Inkson, K. (2007). *Understanding careers: The metaphors of working lives.* Thousand Oaks, California: Sage Publications, Inc.
51. Matte Blanco, I. (1988) *Thinking, feeling and being.* London: Routledge.
52. Humphreys, M., & Brown, A. D. (2002). Narratives of organizational identity and identification: A case study of hegemony and resistance. *Organization Studies, 23*(3), 421–447.
53. Dutton, J. E., Dukerich, J. M., & Harquail, C. V. (1994). Organizational images and member identification. *Administrative Science Quarterly, 39*(2), 239–263.
54. Elsbach, K. D., & Bhattacharya, C. B. (2001). Defining who you are by what you are not: Organizational disidentification and the National Rifle Association. *Organization Science, 12*(4), 393–403.
55. Bowen, D. E., & Ostroff, C. (2004). Understanding HRM–firm performance linkages: The role of the strength of the HRM system. *Academy of Management Journal, 29*(2), 203–221.
56. Nishii, L. H., and Wright, P. M. (2007). Variability within organizations: Implications for strategic human resource management. In *CAHRS Working paper* #07-02. 2007. Ithaca, NY: Cornell University, School of Industrial and Labor Relations, Center for Advanced Human Resource Studies.
57. Nishii, L. H., Lepak, D. P., & Schneider, B. (2008). Employee attributions of the why of HR practices: Their effects on employee attitudes and behaviours and costumer satisfaction. *Personnel Psychology, 61,* 503–545.
58. Grant, A. M., & Hofmann, D. A. (2011). It's not all about me: Motivating hand hygiene among health care professionals by focusing on patients. *Psychological Science, 22*(12), 1494–1499.
59. Bourdieu, P. (1990). *The logic of practice.* Cambridge: Polity Press.
60. Bartlett, F. C. (1995). *Remembering: A study in experimental and social psychology.* Cambridge: Cambridge University Press.
61. Schein, E. H. (1996). Career anchors revisited: Implications for career development in the 21st century. *Academy of Management Executive, 10*(4), 80–88.

62. Rousseau, D. M. (1989). Psychological and implied contracts in organizations. *Employee Responsibilities and Rights Journal, 2*(2), 121–139.
63. Nonaka, I. (1994). The knowledge-creating company. *Harvard Business Review, 69*(6), 96–104.
64. Nonaka, I., & Takeuchi, H. (1995). *The knowledge-creating company*. New York: Oxford University Press.
65. Locke, J. (1988). Two treatises of Government. 1690—republished in 1988, Cambridge: Cambridge University Press.
66. Glaser, B., & Strauss, A. (1967). *The Discovery of grounded theory: Strategies for qualitative research*. New York: Aldine.
67. Charmaz, C. (1990). Discovering chronic illness: Using grounded theory. *Social Science and Medicine, 30*(11), 1161–1172.
68. Charmaz, C. (1994). Between positivism and postmodernism: Implications for methods, In N. K. Denzin (Ed.), *Studies in symbolic interaction*. Greenwich: JAI Press.
69. Henwood, K., & Pidgeon, N. (1995). Grounded theory and psychological research. *The Psychologist, 8*(3), 115–118.
70. King, N. (2004) Using templates in the thematic analysis of text. In G. S. C. Cassell (Ed.), *Essential guide to qualitative methods in organizational research* (pp. 256–270). London: Sage.

Researching HRM to Enhance Understanding: The Neglected Role of Ontology

Brian Harney

Abstract This chapter reviews the dominant methodologies deployed in research examining HRM and its relationship to organisational performance. Having detailed the limitations of dominant scientific approaches the chapter goes on to highlight the merits of explicitly engaging with ontology. Specifically, ontological conversations help in surfacing the role of context and understanding while allowing for emergence and unanticipated consequences. Arguably this leaves researchers better placed to capture the complex and dynamic phenomenon inherent to the employment relationship. The chapter concludes with implications for future research trajectories.

1 Introduction

Research and journals in the domain of work and employment remain largely dominated by quantitative research [1, 2]. Exemplary is Human Resource Management (HRM) where research using a plethora of statistical techniques claims to have demonstrated that a strong link exists between HRM practices and business performance [3, 4]. Less frequent, however, is a sound explanation of *why* or *how* HRM performance impacts take effect [5]. Generally there is failure to fully acknowledge that a simple correlation between two variables constitutes neither a theory nor an explanation [6]. Nonetheless, the solution to the 'explanatory void' [7] in existing studies typically involves a call for greater precision in methodological techniques including better measurement and larger sample sizes [8, 9]. This solution can be likened to a version of Kuhnian 'normal science puzzle solving' [10] whereby the problem is not seen as related to the very assumptions

B. Harney (✉)
Dublin City University Business School, Dublin City University, Dublin 9, Ireland
e-mail: brian.harney@dcu.ie

C. Machado and J. P. Davim (eds.), *Work Organization and Human Resource Management*, Management and Industrial Engineering,
DOI: 10.1007/978-3-319-06376-8_10, © Springer International Publishing Switzerland 2014

deployed, but rather remains something of a methodological conundrum for which a technical solution can be readily found [11]. As a consequence researchers rarely surface and re-examine the key principles that underpin their research, while any method which deviates from the norm or questions the status of the scientific approach is deemed heretical [6, 12]. By contrast this chapter suggests that a move away from a consensus orientation and towards more advanced understanding of work and employment will come from ontological conversations concerning the assumptions that actually underpin research. To date such philosophical introspection has been disappointingly absent in HRM. Yet such neglect may actually be hindering progress in advancing understanding and preventing HRM from accommodating recent changes in the nature and forms through which the employment relationship is managed [7, 13].

2 Ontological Considerations in Researching HRM

This chapter explores the potential of the meta-theoretical vehicle of critical realism (CR) and considerations of social ontology as a middle ground between extreme variants of positivist and interpretivist/phenomenological paradigms. While the term meta-theory is occasionally invoked, it is rarely defined [14]. Here the term is understood in an overarching sense to encompass 'philosophy of social science, ontology, epistemology, methodology and causality' [15]. From the discussion it will be clear that the absence of meta-theoretical discussion in HRM does not equate to philosophical evasion. Indeed, an emerging argument holds that the explanatory void which confronts much research about employment and work may actually be attributable to the very scientific meta-theory that has hitherto underpinned research in this area [7]. In this case viable alternatives will involve more than methodological fine-tuning and answering calls for 'big science' [8], but rather will be found through an exploration of the ontological assumptions that underpin research. As emphasized by Delbridge [16] 'consideration of ontology is fundamental, not optional'.

It has been argued that the dominance of a big science research agenda in HRM research has resulted in a 'narrow, dustbowl empiricism' [17]. In the absence of a robust theoretical underpinning HRM, and its performative version in particular, is often exposed as something of a naked emperor with statistical convenience/sophistication coming at the expense of pragmatic relevance and theoretical reflection [18, 19]. Moreover, in interpreting large quantitative studies it is perhaps too easily forgotten that a statistical association implies correlation not causation [20]. Within HRM Kane [21] sounded early caution noting that 'the development and evaluation of constructs and theories has been subordinated to- and to some extent eclipsed by- the search for lawful relationships among empirically observed phenomena'. Indeed this is the basis of the critique that Kaufman [22] directs at HRM, arguing that extant approaches have perpetuated a limited domain and misspecified supply side understanding of the concept. Instead, Kaufman holds that the focus of research attention should be on the demand side, exploring the pattern of HRM adoption among firms. The problem with Kaufman's solution, however, is that leaning back on the crutch of mainstream economics is unlikely to address the problems of abstractness and simplification. Thus, while large-scale quantitative research in HRM has been useful in capturing the extent of phenomenon its ability to capture process and context in a theoretically informed way is limited. In light of these arguments it is perhaps unsurprising that researchers still find it necessary to call for more small-scale contextual studies of HRM [23, 24].

In advancing the meta-theoretical conversation critical realism eschews the positivist search for universal laws and instead seeks to identify 'tendencies' associated with the operation of structural mechanisms in open systems [25]. Critical realism charges dominant scientific research with committing the 'epistemic fallacy'; letting questions of what we can know/how we may have knowledge (epistemology) determine what can exist/is knowable [26, 27]. A critical realist under-labourer offers a way to overcome the epistemic fallacy by capturing generative mechanisms and emergence [27, 28]. Explicit consideration of ontology therefore surfaces a range of philosophical concerns which have been muted within HRM, partly because the assumption of theory neutral observation mitigates against critical reflection [12]. At the analytical heart of critical realism lies 'the dynamic relationship between the generative potential inherent in social structures and its contingent realization through corporate agency, as well a form of causal explanation sensitive to contextual complexity' [29]. If the relationship between HRM and Performance is judged to be complex, uneven and variable, critical realism would seem to offer one potential for contextually rich explanatory accounts of HRM interventions [29]. For the present purposes, the next sections consider five main 'domain assumptions' associated with social ontology. This illuminates the value of engaging on ontological conversation and serves as a way to appreciate the prospective limitations of the dominant 'scientific paradigm' in HRM performance research and to open up alternative perspectives.

(1) Realist versus Empiricist Ontology

Realism denotes the assumption that an external reality exists of things not observable or measurable beyond our senses which can nevertheless be real and known. As Johnson and Duberly note, such intransitive objects exist and can act independently of their identification in human knowledge; this is Bhaskar's [26] 'intransitive domain'. The 'transitive domain', on the other hand consists of socially produced knowledge that facilitates us in making sense of the world, largely through theoretically determined descriptions. Purcell and Kinnie have hinted at the importance of such insights in HRM; 'methodological considerations have determined what questions have been asked while factors beyond the reach of the chosen method, however important, have been ignored' [31]. To follow dominant scientific logic is to suggest that simply because an empirical association or regularity does not exist that there is no causal connection. However, this is 'misleading in the extreme' [32]. Critical realism on the other hand offers a route which overcomes this limitation of by going beyond surface appearances to establish a theoretical understanding of the causal powers of real structures and mechanisms that account for the reality which we observe [20]. Hence the notion of an exclusive empirical world existing only through observations and direct verifiability is supplanted by a deeper realist ontology. Notably, the idea of a reality totally independent of our representations is purely an ontological one, it is not an epistemological position on how we know reality [33]. This does not imply a degeneration into the chimera of 'judgemental relativism' where there are no grounds for preferring one knowledge claim to another. Instead, critical realists invoke the notion of epistemic relativism to consider the 'pragmatic adequacy' or meaning of a theory in relation to the objects to which it refers [30]. Such understanding is one reason why the notion of complexity is said to sit well with the epistemology of critical realism.

(2) Open versus Closed Systems

The distinction between open and closed systems is central to critical realism. Critical realism takes issue with the actualist ontology underpinning dominant scientific aproaches. An actualist ontology is defined in terms of a closed system logic made up of a fixed set of unchanging regularities which are judged to be manifestations of universal causal laws. This is the epistemic fallacy in action, 'given an empiricist epistemology the ontology produced is an actualist closed systems model of being' [34]. This logic is diffused into HRM research as 'to suggest that some HRM practices are statistically associated with increased performance, is to assume regularity and hence closure' [35]. However, the open systems of the social world mean that such regularity and closure are rare even in the natural realm [36]. Specifically, explicating Lawson's critique of empirical realism to the domain of HRM suggests that to assume that HRM can automatically be reduced to productivity enhancing or damaging effects which are realized as intended is to commit to the *fallacy of atomism*, while to assume that HRM operates in isolation from the institutional context and broader dynamics of the economy in which is embedded is to commit to the *fallacy of isolationism*. It has long been noted that relations at the point of production are complex and variable [37]. Critical realism holds that the social world operates as an open system so that the regularity and closure presupposed for universal causal laws are 'more or less non-existent' [36].

(3) Layered versus Flat ontology

A third key domain assumption involves a stratified ontology. As Edwards [38] argued in his earlier work 'the metaphor of levels helps in dealing with the long-standing analytical problem of relating structure to action'. For Bhaskar [39] a layered ontology is made up of three dimensions: (1) the real, consisting of causal powers, enduring structures, generative mechanisms and their tendencies. These 'real' non empirical structures and mechanisms can produce (2) the actual, consisting of events, entities, and/or states of affairs, only some of which are conceptually mediated as (3) the empirical, consisting of experiences, perceptions and impressions of events. In contra to positivism and interpretivism, critical realism's layered ontology can comprehend emergence, tension and discontinuities. This allows causal mechanisms to exist even when regularities fail to obtain (vs. positivism), while they can also be understood to operate irrespective of their direct identification by agents (vs. interpretivism). It is this multi-layered ontology which differentiates critical realism from extreme interpretative accounts which see the world in a constant state of flux. Instead, critical realism considers issues such as why construction takes a particular form under given conditions thereby offering a genuine attempt to understand contextual embeddedness [40]. Critical realism proposes that there are deeper underlying structures and generative mechanisms ‘lying beneath the surface phenomena of direct experience’ [41]. The stratification between the levels of the empirical and the real allows for a disjuncture between underlying causal mechanisms and observable phenomenon [42]. Thus, critical realism offers a genuine attempt to understand institutional embeddedness and provide contextually rich accounts. Possibilities can exist which are never actualized, such *real* yet *unactualized* potentialities remain as *tendencies* in the deeper layers of reality. In contrast to an actualist ontology, if correlations are not observed, this does not mean that causal mechanisms does not act in regular ways, only that its effects are on occasion suppressed by the action of other causal mechanisms.

(4) Explanation versus Prediction

The danger of an exclusive reliance on quantitative methods is that the desire for prediction takes precedent over developing sufficient explanation or understanding. This can be seen in HRM-Performance studies commencing with the work of Huselid [43] who asserted with great precision that ‘a one standard-deviation increase in High Performance Work Practices is associated with a relative 7.05 % decrease in turnover and, on a per employee basis, $27,044 more in sales and $18,641 and $3,814 more in market value and profits, respectively’. Studies have since attempted to make this argument on a cumulative basis. Combs and colleagues [4] reviewed data from 92 studies covering a total of 19,319 organisations noting that 20 % of the utility available from predicting performance differences amongst organisations was attributable to High Performance Work Systems (HPWS). Nonetheless it remains the case that much remains unknown regarding the myriad of mechanisms through which performance is altered or affected by a firm’s HR system [44].

In contrast the domain assumptions of an open system and a layered ontology associated with critical realism help ground the logic that the essence of social science is not the prediction of event regularities but rather explanation. Hence, critical realist investigations merely identify ‘tendencies’ associated with the operation of structural mechanisms in open systems [25]. From this platform it is recognised that the contexts in which mechanisms operate will determine whether their impact is realised and the contingencies influencing this process [20]. This in turn opens up the possibility to ‘bring out the complexity of phenomenon’ [45]. In HRM terms this means studies into the boundary conditions shaping the content and process of HRM practices e.g. critical contingencies such as strategy and industry sector coupled with greater employee assessment of HRM [46]. The mode of inference then becomes retroductive, asking what the world must be like for such conditions to exist, with analysis working backwards to gain a theoretical understanding of the mechanisms of cause and effect which underlie empirical tendencies. Although these are not directly observable, they underlie and govern the events of experience and hence explain *why* regularities occur [30]. For example, the employer-employee relationship itself presupposes the existence of capitalist structures and activities e.g. contract, wages, hiring and firing and so forth. Here theory is indispensable to facilitate understanding and in order to provide richer causal explanations.

The argument that causal explanations have to identify underlying mechanisms would seem quite useful in addressing some of the dilemmas facing the HRM field. For example, the Resource Based View (RBV) of the firm views the employment systems as a complex social structure. Central concepts include *causal ambiguity* as subtle interconnections between contingent factors and *path dependency* which recognizes the emergent nature of strategy, and HR decisions based on what 'feels' right and fits with an organisation's culture. Despite the fact that this complexity is continuously put forward as the basis for making each organisation's experience unique and thereby offers a route to sustainable competitive advantage it is 'routinely overlooked' [6]. Further, and very much in keeping with the logic of the RBV, critical realism helps capture the essence of entities existing independent of their articulable identification. This might include tacit knowledge and cultural effects. Unfortunately, however, HRM Performance research still craves the fixed regularities, closed systems of classical scientific approaches, and so our understanding remains narrow and partial [12].

(5) Multiple Methods

Ontological consideration serves to prevent premature methodological foreclosure as one method of study is not favoured or privileged over another. While the dominant scientific approach of quantitative methods has been subject of the harshest criticism this is only as result of their application under the tenets of logical positivist assumptions [47]. Instead those who explore social ontology and critical realist logic embrace a multi-method approach, recognizing the potential that qualitative analysis can bring to bear on surfacing complexities, processes and contextual idiosyncrasies. Ragin's [48] thesis that it is possible to combine case and variable orientated research, in particular by identifying sets of causal conditions that may affect some outcome, may have particular value here. This argument for multi-methods for HRM-Performance research is not novel, however critical realism ensures this call is underpinned by an appropriate philosophical under-labourer.

In enforcing this point it is useful to draw on Sayer's [49] distinction between extensive and intensive research designs. Extensive approaches are those that aim to provide information on the general properties, prevalence and distribution of a phenomenon in a broad population, while intensive research designs are those that explore how some causal processes work in a limited number of instances.

Ragin [48] makes a somewhat similar distinction between what he terms 'variable' and 'case' orientated research strategies respectively. The approach advocated here is essentially pluralist in that it legitimates both. The key point is that in both instances authors are pragmatic in understanding subsequent implications for methods. Surveys, for example, need not be devoid of attempts to understand the social construction of meaning and can be embedded as part of case study research, while is no a priori reason why case studies should be considered more exploratory and illustrative and surveys more explanatory. Indeed, the very distinction between exploratory and explanation is judged rather ambiguous [50].

3 Discussion

This chapter has presented an overview of HRM-performance research and suggested the benefits of deeper ontological conversations. The limits of relying exclusively on a 'big science' approach to HRM stems from its quest for surface level empirical regularities. By contrast greater ontological depth offers one way to better accommodate the issues of complexity and context that characterise the employment relationship. The utility of a stratified ontology in accommodating these realities stems from its depth and focus on inert causal capacities and their contingent operation as tendencies at one layer that may *condition* but cannot be *collapsed* indeterminately to explain realities at another. This type of understanding exposes the limitations of '*big science*' Humean causality and its cornerstone of one dimensional constant conjunctions and instead points to the explanatory promise of '*complex causality*' [51] and its project of explaining the observed features of the real world by reference to the underlying mechanisms that are causally responsible for them [52]. It has been long noted that a complete understanding of the employment relationship requires that we analyse the social relations that underpin them.

The dualism of structure and agency that forms the analytical core of a critical realist approach mirrors that found in the open systems logic and more integrative, multi-level frameworks of HRM [2]. In so doing there is a recognition of the importance of capturing the totality of HRM interventions while also emphasizing that *how* structural influences become manifest is contingent on managerial and employee interpretation and intent. Appreciation of the interaction between structure and agency should form a key component in enhancing our understanding of HRM. At the same time, it can accommodate the reality that HR practices and rules have to be interpreted in action; the indeterminacy of the employment relationship holds that this enactment process is subject to constant dynamic of 'conflicted collaboration' [53].

Evidently a social ontology approach is not without its limitations and critiques. Contu and Wilmott [54] contend that there is a tendency to privilege structure at the expense of agency. For those with a penchant for classical modes of inquiry the approach does not lend itself to the development of clear hypothesis, while with a focus on understanding a critical realist informed approach is somewhat backward looking rather than predictionist *per se* [36, 40]. It follows that theoretical explanations are not judged either correct or false, nor assessed by reference to a narrow quest for empirical consistency. Instead, the central concern is the plausibility of theory and its analytical utility as 'a 'point of entry' for the analysis and understanding of the phenomenon in question' [47]. Moreover, theoretical understanding is located in the *transitive realm* where interpretations will always be fallible and open to contestation and revision. It is here, in arguments relating to epistemology, that criticism of a critical realist informed approach finds its strongest force. As Halfpenny [55] notes 'the problem of finding grounds on which to justify the existence of underlying causal mechanisms is the epistemological stumbling block of realism, just as the problem of causation is of positivism and the problem of interpretation is of interpretivism'.

Overall, sympathy for a critical realist inspired inquiry ensures that explanatory accounts are privileged, that *processes* of HRM are adequately contextualised, and that informality and unanticipated consequences can be accommodated. This is matched by the promise of avoiding the limitations of the big science agenda without becoming lost in the chimera of extreme relativism [55]. Critical realist understanding begins with a valuable exercise of immanent critique which exposes the problems with extant accounts and by so doing sets in motion the search for alternative conceptual resources which might enable these shortcomings to be overcome [27]. Interrogation and critical questioning of this type are strangely muted in extant work on HRM. This is especially the case in HRM-performance studies. Ultimately recognition of structural layers and emergence leaves scope for more critical understanding. Further, because critical realism holds that theories are always fallible in the transitive domain and so absolute closure is not possible, it affords the opportunity for reassessment and critique. This provides scope for much needed on-going critical dialogue, encouraging sensitivity to contradictory pressures rather than simply proffering immediate, universal solutions. Applying this logic means that HRM may be treated as something that managers *do*, albeit conditioned by external determinants and internal dynamics, as opposed to something an organisation merely *has*. Specifically, by focusing attention on generative causal mechanisms ontological considerations offer the potential to provide theoretically informed, contextually rich, explanatory accounts of HRM interventions. In following this road less travelled HRM researchers might be better placed to consider emergent processes, outcomes and new organizational forms.

4 Conclusion

Overall, sympathy with a critical realist meta-theoretical underpinning and the contextual and theoretical sensitivity that it provides for exploring HRM should not be particularly contentious. Edwards contends that critical realism simply 'provides a grounding for what industrial relations researchers have often tacitly done' [56]. There is also a notable overlap between the general argument advocated in this chapter and Watson's [57] call for critical social science in HRM in that both stress the importance of (a) understanding actual HRM processes in '*real*' organisations (b) exploring the relationship and tensions between HR activities and patterns of culture, power and inequality, (c) theorising HRM activities in such a way that full weight is given to both, human/managerial agency and to structural circumstances, and d) appreciating the interaction between managerial agency/choice and structural constraints and opportunities in the shaping of HRM practices [7, 57]. In sum, critical realism privileges ontology, emphasising the way the world actually is as opposed to committing the epistemic fallacy of 'assuming a certain world order and then mapping reality back to fit this order' [7]. This may very well offer a vehicle for HRM researchers to enhance their understanding by supporting more holistic and contextually embedded research approaches.

References

1. Batt, R., & Banerjee, M. (2012). The scope and trajectory of strategic HR research: Evidence from American and British Journals. *International Journal of Human Resource Management, 23*(9), 1739–1762.
2. Jiang, K., Lepak, D., Hu, J., & Baer, J. (2012). How does human resource management influence organizaional outcomes? A meta-analytic investigation of mediating mechanisms. *Academy of Management Journal, 55*(6), 1264–1294.
3. Fleetwood, S., & Hesketh, A. (2010). *Explaining the performance of human resource management*. Cambridge: Cambridge University Press.
4. Combs, J. G., Ketchen, D., Hall, A., & Liu, Y. (2006). Do high performance work practices matter? A meta-analysis of their effects on organisational performance. *Personnel Psychology, 59*, 501–528.
5. Harney, B., & Jordan, C. (2008). Unlocking the black box: Line managers and HRM performance in a call centre context. *International Journal of Productivity and Performance Management, 57*(4), 275–296.
6. Hesketh, A., & Fleetwood, S. (2006). Beyond measuring the HRM-organisational performance link: Applying critical realist meta-theory. *Organization, 13*(5), 677–699.
7. Harney, B. (2009). Exploring the road less travelled in HRM-Performance research: A critical realist alternative to big science. Proceedings of the Labor and Employment Relations Association 61st Annual Meeting LERA.
8. Wall, T., & Wood, S. (2005). The romance of human resource management and business performance, and the case for big science. *Human Relations, 58*(4), 429–462.
9. Wright, P., & Haggerty, J. (2005). Missing variables in theories of strategic human resource management: Times, case and individuals. *Management Review, 16*(2), 164–173.
10. Kuhn, T. S. (1962). *The structure of scientific revolutions*. Chicago: The University of Chicago Press.
11. Kincaid, H. (1996). *Philosophical foundations of the social sciences*. Cambridge: Cambridge University Press.
12. Dipboye, R. (2007). Eight outrageous statements about HR science. *Human Resource Management Review, 17*(2), 96–106.
13. Fleetwood, S., & Hesketh, A. (2006). HRM-performance research: Under-theorized and lacking explanatory power. *International Journal of Human Resource Management, 17*(12), 1977–1993.

14. Tsoukas, H. (1994). What is management? An outline of a metatheory. *British Journal of Management, 5*, 289–301.
15. Fleetwood, S., & Hesketh, A. (2008). Theorising under-theorisation in research on the HRM-Performance Link. *Personnel Review, 37*(2), 126–144 (p. 12).
16. Delbridge, R. (2006). Extended review: The vitality of labour process analysis. *Organization Studies, 27*(8): 1209–1219 (p. 1210).
17. Dipboye, R. (2007). Eight outrageous statements about HR science. *Human Resource Management Review, 17*(2), 96–106 (p. 100).
18. Guest, D. (1997). Human resource management and performance a review and research agenda. *International Journal of Human Resource Management, 8*(3), 263–276.
19. Guest, D. (2011). Human resource management and performance: Still searching for some answers. *Human Resource Management Journal, 21*(1), 3–13.
20. Godard, J. (1989). Beyond Empiricism: Alternative philosophies of science and the study of industrial relations. *Queen's Papers in Industrial Relations, 3*(March), 1–46.
21. Kane, J. (1991). Toward a modernized model of science. *Human Resource Management Review, 1*(4), 245–251 (p. 245).
22. Kaufman, B. (2010). A theory of the firm's demand for HRM practices. *International Journal of Human Resource Management, 21*(5), 615–636.
23. Janssens, M., & Steyaert, C. (2009). Point-counterpoint HRM and performance: A plea for reflexivity in HRM studies. *Journal of Management Studies, 46*(1), 143–155.
24. Paauwe, J. (2009). HRM and performance: Achievements, methodological issues and prospects. *Journal of Management Studies, 46*(1), 129–152.
25. Thompson, P. (2004). Brands, boundaries and bandwagons. In S. Fleetwood & S. Ackroyd (Eds.), *Critical realist applications in organisation and management studies*. London: Routledge.
26. Bhaskar, R. (1978). On the possibility of social scientific knowledge and the limits of naturalism. *Journal for the Theory of Social Behaviour, 8*(1), 225–252.
27. Cruickshank, J. (2003). *Realism and sociology: Anti-foundationalism, ontology and social research*. London: Routledge.
28. Sayer, A. (2004). Forward: Why critical realism. In S. Fleetwood & S. Ackroyd (Eds.), *Critical realist applications in organisation and management studies* (pp. 6–21). London: Routledge.
29. Reed, M. (2005). Reflections on the 'realist turn' in organization and management studies. *Journal of Management Studies, 42*(8), 1621–1644 (p. 1637).
30. Johnson, P., & Duberley, J. (2005). *Understanding management research*. London: Sage.
31. Purcell, J., & Hutchinson, S. (2007). Front-line managers as agents in the HRM-performance causal chain: theory, analysis and evidence. *Human Resource Management Journal, 17*(1), 3–20 (p. 540).
32. Hesketh, A., & Fleetwood, S. (2006). Beyond measuring the HRM-organisational performance link: Applying critical realist meta-theory. *Organization, 13*(5), 677–699 (p. 678).
33. Searle, J. (1995). *The social construction of reality*. London: Allen Lane.
34. Cruickshank, J. (2003). *Realism and sociology: Anti-foundationalism, ontology and social research*. London: Routledge (p. 97).
35. Hesketh, A., & Fleetwood, S. (2006). Beyond measuring the HRM-organisational performance link: Applying critical realist meta-theory. *Organization, 13*(5), 677–699.
36. Lawson, T. (2003). *Reorienting economics*. London: Routledge.
37. Edwards, P., & Wright, M. (2001). High-involvement systems and performance outcomes: the strength of variable, contingent and context-bound relationships. *International Journal of Human Resource Management, 12*(4), 568–585.
38. Edwards, P. (1986). *Conflict at work: A materialist analysis of workplace relations*. Oxford: Blackwell (p. 60).
39. Bhaskar, R. (1998). *The possibility of naturalism: A philosophical critique of the contemporary human sciences*. London: Routledge.

40. Edwards, P. (2006). Industrial relations and critical realism: Industrial relation's tacit contribution. *Warwick Papers in Industrial Relations, 80*(March), 1–20.
41. Lawson, T. (2003). *Reorienting economics*. London: Routledge (p. 229).
42. Steinmetz, G. (2004). Odious comparisons: Incommensurability, the case study, and the "small N's" in Sociology. *Sociological theory, 22*(3), 371–400.
43. Huselid, M. A. (1995). The impact of human resource practices on turnover, productivity, and corporate financial performance. *Academy of Management Journal, 38*(3), 635–672 (p. 667).
44. Patel, P., Messersmith, J., & Lepak, D. (2013). Walking the tightrope: An asessment of the relationship between high-performance work systems and organizational ambidexterity. *Academy of Management Journal, 56*(5), 1420–1442.
45. Outhwaite, W. (1983). Towards a realist perspective. In G. Morgan (Ed.), *Beyond method: Strategies for social research*: 321–330. London: Sage Publications (p. 328).
46. Jiang, K., Takeuchi, R., & Lepak, D. (2013). Where do we go from here? New perspectives on the black boxes on strategic human resource management research. *Journal of Management Studies, 50*(8).
47. Godard, J. (1993). Theory and method in industrial relations. In R. J. Adams & N. M. Meltz (Eds.), *Industrial relations theory: Its nature scope and pedagogy* (pp. 283–306). Metuchen: IMLR Press and Scarecrow.
48. Ragin, C. C. (1987). *The comparative method: Moving beyond qualitative and quantitative strategies*. California: University of California Press.
49. Sayer, A. (1992). *Method in social science: A realist approach* (2nd ed.). London: Routledge.
50. Ryan, B., Scapens, R., & Theobold, M. (2002). *Research Method and Methodology in Finance and Accounting* (Second Edition ed.): Thomson.
51. Fiss, P. (2007). A set-theoretic approach to organizational configurations. *Academy of Management Review, 32*(4), 1180–1198.
52. Fleetwood, S., & Hesketh, A. (2008). Theorising under-theorisation in research on the HRM-Performance Link. *Personnel Review, 37*(2), 126–144.
53. Delbridge, R. (2007). Explaining conflicted collaboration: A critical realist approach to hegemony. *Organization Studies, 28*(9), 1347–1357.
54. Contu, A., & Wilmott, H. (2005). You spin me around: The realist turn in management studies. *Journal of Management Studies, 42*(8), 1645–1662.
55. Halfpenny, P. (1987). Laws, causality and statistics: Positivism, interpretivism and realism. *Sociological Theory, 5*, 33–36 (p. 35).
56. Edwards, P. (2006). Industrial relations and critical realism: Industrial relation's tacit contribution. *Warwick Papers in Industrial Relations, 80*(March), 1–20 (p. 6).
57. Watson, T. (2004). HRM and critical social science analysis. *Journal of Management Studies, 41*(3), 447–467.

Examining the Motivation to Work Within a Group of Gardeners with Mental Disorders

Ester Machado Pereira and Carolina Feliciana Machado

Abstract This paper has examined the motivation to work of a group of gardeners with mental disorders. The aim of this study was to identify the level of motivation of each mentally disturbed worker and to apply motivation theories to everyone in the group. The results obtained have shown how these different theories were used to examine the motivation of the workers and it also allowed us to make suggestions to raise their motivation.

1 Introduction

In 1999 (just at the end of twenty century), 1.500 million people existed with mental disorders [1] and this number is on the rise. Indeed, at the present, and according to CCHR International [2] "120 million people worldwide have been diagnosed with mental disorders". This is reason enough to concern us with this special kind of person. Not only the psychologists or psychiatrists, but also human resources managers should be concerned because mental patients work within many organisations. Human resources managers have to consider this fact in their staff evaluations. Therefore it is of great need to increase research in this area and also to analyse mental patients in their working context.

In our study we analysed the motivation of a group that has mental disorders. It is a very important task to examine motivation because on the one hand, it has

E. M. Pereira
School of Economics and Management, University of Minho, Campus Gualtar, 4710-057 Braga, Portugal

C. F. Machado (✉)
Department of Management, School of Economics and Management, University of Minho, Campus Gualtar, 4710-057 Braga, Portugal
e-mail: carolina@eeg.uminho.pt

C. Machado and J. P. Davim (eds.), *Work Organization and Human Resource Management*, Management and Industrial Engineering,
DOI: 10.1007/978-3-319-06376-8_11,

direct relation to organizational efficiency, and on the other hand, it is a difficult task because you cannot see motivation. What we can observe though is multiple behaviours and the results of these behaviours.

The aim of this study was to identify the level of motivation of each mental patient and to apply motivation theories for every worker in this group. We suggested a number of important practical implications of this study.

2 Theoretical Background

Most of the theories of work motivation presuppose that "the individual's behaviour seeks to increase their own personal satisfaction and to neutralize the dissatisfaction factors" [3, p. 174].

Many motivation theories exist and all of them have strong and weak aspects. No model is perfect, but each one adds something new to our understanding of motivation.

Our task has consisted in choosing several models that could be applied in our group of gardeners.

We kept in mind the following points:

- *The working place:* it is a store, where they actually handle the flowers. It is not a factory with an assembly line; if this was the case, we could opt for specific models for that type of organisation.
- *The type of people:* as we have previously mentioned we worked with mental patients. These people require special treatment, and not every incentive is adopted.
- *The limitations of the administration:* we found models, which ones the administration has had difficulties to implement; these models were left on the side.

We also paid attention to ensure that the models we selected were diverse. They are mentioned below:

- Theory X and Theory Y of Douglas McGregor
- The ERG theory of Alderfer
- The Hackman and Oldham model
- The expectancy theory of Lawler.

The last two models are satisfaction models. The difference between motivation theories and satisfaction theories is explained as follow: "The motivation models try to explain the behaviours through the study of variables, which consider the individual and the situation which the behaviour has developed" [3, p. 174]. By its way, "The satisfaction models try to measure the quality of life at work and are related to emotional states that represent an affectionate response which results from workers' experiences in relation to the work" [4, p. 59].

On the one hand, the motivation could be connected with satisfaction. A person, who is motivated, could feel satisfied. On the other hand, empirical research demonstrates that there is no casual relationship between the two concepts.

"The idea that ≪a satisfied employee is a productive employee≫ has long time been a strong conviction of management, but it does not correspond to reality" [4, p. 59].

For a better understanding of our study, we are going to explain each model concisely.

In relation to Theory X and Theory Y of McGregor, this author says that individuals can be divided in two according classes with their attitudes and behaviours, in other words, Theory X and Theory Y [5]. Theory X represents the assumptions on which traditional organisations are based. In that theory, McGregor affirms that the human being has a natural aversion to work and avoids it whenever it is possible.

The majority of people must be coerced, controlled, directed and threatened with punishments to meet the objectives of the company. An individual can be motivated through physiological and security levels.

On the contrary, in Theory Y, McGregor declares that physical and mental efforts in the working place are as natural as those that one uses in his/her hobbies. Punishments are not the only way to meet the objectives of the organisation. People become attached and committed to the objectives of their organisation through means of self-control and self-direction. Individuals like to assume responsibilities and use creativity to solve organisational problems. The motivation occurs at the connection of esteem and self-actualisation levels as well as the physiological and security levels. McGregor implies that a Theory Y approach is the best way to elicit cooperation from members of an organisation [5, 6].

It depends on an individual's characteristics as to whether Theory X or Theory Y is applied. They are mutually exclusive.

It is the task of management to create conditions in which persons may satisfy their motivational needs so that they can achieve the goals of the organisation and their own.

In regard to the theory of Alderfer, this author presents three categories: existence needs that are basic needs; physiological and safety needs of a material nature [7]. In managerial terms, it corresponds to the base wage and material benefits.

The relatedness needs include the interpersonal relationships of an esteem nature in the working place.

Finally, the growth needs, where the individuals can develop their potential and create self-esteem and self-actualisation.

Alderfer suggests that individual's progress through the hierarchy; from existence needs, to relatedness needs, to growth needs, as the lower level needs become satisfied.

This author makes three basic propositions relating to the three-need relationship [3, 7]:

1. When one level is satisfied the individual activates desire for the next level.
2. The more a level is satisfied the more it will be desired.
3. The lower level needs become the main focus of the individual's efforts when the individual cannot be satisfied with the level which is above.

The authors of the Hackman and Oldham model explain how the relationship is processed between the content of the work and the motivation. They propose a list of characteristics of the work. These characteristics will be able to influence the motivation. The five dimensions of the list are [8–10]:

- *Skill variety:* increasing the variety of the work is a form of fighting against the boredom; consequently, Hackman and Oldham appeal to the enrichment of functions to combat the monotony.
- *Task significance:* to give a sense of meaning to the work, show that the work demonstrates a social value.
- *Task identity:* the individual has the possibility of identifying the results of his/her work.
- *The autonomy:* the independence and the freedom that the individual has to choose the procedures to use in his/her work, in accordance with the objectives of the company.
- *The information:* the worker needs to have access to precise, comprehensible and frequent information about the quality of his/her work.

Relative to the theory of expectancy of Lawler (1971), he affirms that money can motivate the acting, but also the companionship, and the dedication. However, a majority of organisations do not apply the sufficient payment to motivate their workers [11–13]. Reasons for that fact can be [14]: increasing of salary that are given regularly and too modest; they can give the impression that their gains are independent of their performance. Besides, managers make no distinction in the evaluations of performance, especially if evaluations originate different salaries. Perhaps the comparisons with valued people could bring problems to the relationship. Another reason can be related to the fact that the policy of remuneration of the organisations is connected to governmental policies, which try to regulate the salaries in order to combat inflation. Finally, the prejudice created by the Human Relation Approach in regard to the salary—which defended that the workers are motivated principally by needs of acknowledgement, social approval and participation [15], as well as the existence of the psychological limitations of the Scientific Administration of F. W. Taylor—who described that the workers were only motivated by economic incentives.

In regard to the systems of payment, he recommends that it should be individualized; it has to be adjusted to the needs of the organisation and of the workers.

In short, the Theory X and the Theory Y of Douglas McGregor shows us two types of individuals: those who do not like to work and are motivated by punishments and threats to fulfil their functions. These workers need a supervision which directs and controls their labour (Theory X). On the contrary, the Theory Y mentions that the individuals face their job as natural. The employees are motivated through their realisation in their work. They assume responsibilities and creativity.

Relative to the ERG theory of Alderfer, this author suggests three categories: existence, relatedness and growth needs. The individuals can pass to the next category when they have satisfied the present level.

Hackman and Oldham [16] propose in their model, a list of aspects which have an effort on the motivation. These characteristics are: skill variety, task significance, task identity, autonomy and information.

In relation to the expectancy theory of Lawler he gives to the payment a great significance. Money influences the motivation of the workers but, unfortunately, managers can not make use of this factor.

3 Methodology

The technique used to analyse the reasoning of mental patients, was an individual interview that was semi-directed. Within this method the interviewer uses a series of open-ended questions that aren't necessarily asked in the same order as they were annotated. This is important because mental patients cannot always have a coherent conversation and the interviewer has the flexibility to change the order of the subjects, or ask new questions. Because of this the interview can be more flexible and the interviewer can acquire a greater amount of data. The interviews were recorded and transcribed.

It was also elaborated on an inquiry that contained 21 sentences, some that aimed at high motivation and the others at low motivation. The researchers filled out the inquiry so that the rate of answers was maximized. The reason for the inquiry was to control the answers that were given in the interview and to analyse if there were or were not contradictions between them.

In relation to the content of the interview and the inquiry, the following areas were included: the working place, the relationship between the elements of the group, the instructor and the organisation where they worked.

The universe of the group was a man and seven women with ages between 21 and 46 years. Most of them had low qualifications. They were working at a Health Centre as gardeners.

The members of the group had the following mental disorders: three with schizophrenia, two with personality disorders, one with epilepsy, one mentally disturbed and one with chronic alcoholism. Of these people, three lived at the Health Centre; the others lived with their families or in their own apartment.

When the interviews and the inquiries of the mental patients were evaluated, the following approaches were considered.

In relation to the workers, we kept these items in mind:

- whether they understood the questions or not
- the "quality" of the answers of the individual; if the questions were answered directly, or if the workers were imprecise in their answers
- eventual contradictions in the data mentioned in the interview and/or in the inquiry.

The following, abbreviated, list was elaborated on in order to measure the degree of each worker's motivation.

Points that revealed:	
A high motivation	A low motivation
• Enjoyment of certain types of tasks	• Dissatisfaction of doing certain tasks
• Suggestions to modify the working place	• Critiques that were exaggerated in order to improve the working conditions
• Satisfaction of receiving a fair salary	• Not satisfied wages
• Steadiness of the position in the working place	• Unjustified deficiencies
• Good relationship with the other members of the group	• Conflicts between the workers and the isolation from the group
• Good relationship with the instructor	• Misunderstandings and conflicts with the instructor
• Incentives and praises received from the instructor	• Critiques and reprehensions received from the instructor
• Statements in the inquiry that revealed a high motivation	• Statements in the inquiry that pointed out a low motivation

Special attention was paid to the following items in the interview:

- the alteration of the order of the questions (an excessive change of the questions could signify that the workers did not answer the subjects directly)
- the techniques used in the interview (for example, to motivate the interviewee to answer the questions)
- introducing new subjects in the interview
- the difficulties felt in the interview and in the inquiry.

4 Results and Discussion

First, a report on how the interviews and inquiries went is going to be given. Some difficulties were faced: occasionally the interviewees gave long answers; researchers did not interrupt them because they could supply important data; or there were also workers, who had to be stimulated to speak. Whenever it was appropriate, researchers changed the order of the questions or formulated new subjects depending on the context of the interview.

Finally it were detected only a few contradictions between the interview and the inquiry.

After analysing the interviews and the inquiries, it was detected that 50 % of the workers had a high motivation (on a scale from "no motivation" up to "very high motivation").

The other people were motivated moderately and only one element was not motivated in her working place. This was a good result, however we feel we can elevate their motivation by proposing measures that can help them feel better about gardening.

In regard to the theories we have concluded the following:

In relation to Theory X: all the workers liked to be guided because they didn't always understand exactly what they were supposed to be doing.

> We are encouraged to work as alone as possible, that means, without any direction from the instructor. I think this is a contradiction because we receive orders and directives that only the instructor can give us, and sometimes we have another problem: everyone gives orders and no one is in charge. (man, external)

They also like to feel secure.

It should be to reject the coercion, threats, and punishments, so the individual is not pressured because they have difficulties working with psychological tensions.

Researchers also do not agree that people are only motivated by economic incentives, reason why more alternatives will be mentioned later. Due to their pathology the workers need the support of their organisation. For example, they should be administered, motivated and controlled by the same. But it is almost impossible to control the unexpected in the organisation.

Theory Y has more application in this group. For these individuals the work is like therapy: it makes them feel good, useful, accomplished and the relationships within the group are very important to them:

> When I'm feeling good, I like to work on my own, but when I'm not feeling good, I like to work in the group. (woman, internal)

In addition, they would like to assume more responsibilities and they have ideas to improve their working place.

However, we cannot apply all presuppositions. For example, the individuals have difficulties orientating themselves, using their imagination, and solving problems, because of their pathology. Another reason is they do not assume more responsibility in their working place because it depends on their pathology that does not permit them to execute tasks more responsibly. Besides, they feel safer in being guided by their instructor.

In conclusion, some useful suggestions can be presented in order to improve the work of these mental patients:

- to develop the potential of each person (we described the tasks that were most suitable for each worker) in order to reach the goals of the organisation
- to create new operation methods that were presented below in the application of the Hackman and Oldham model
- to decentralize and to delegate; we have suggested which workers could assume more responsibility in the group
- to participate in decisions of the organisation; we have presented the workers which show interest in participating in the decisions of the management
- evaluation of the work of each mental patient by the instructor: this is very important because they have to know if they are working correctly or not due to their own difficulties in self examination.

In relation to the ERG theory of Alderfer, most of the mental patients are in the second level: relatedness needs. One of the reasons for this is that the internal mental patients have their existential needs satisfied primarily by the Health Centre. Only two members of the group are in the primary level because their existential needs were not satisfied.

The persons who do not reside in this organisation have their existential needs partly satisfied. In general, it is the only income that they receive and it is not enough for satisfy all their living needs. Nevertheless, they are also in the second level. In the interviews they mentioned that they help each other. This is very significant, especially when the mental patients are in a bad mood. Within the group they get encouragement and support from their co-workers.

> When somebody is feeling bad, normally their colleagues go to her and give her encouragement, except when the person rejects the help… (man, external)

The workers are gardeners not only to accomplish the requested tasks, but also to feel integrated and to know how to work as a team due to the changes of dispositions in their pathology. These obstacles can influence the performance of their work, the relationship between the members of the group and, at last, their emotional balance.

The majority of the mental patients have a great need to communicate within the group.

> It is obvious that we don't come here to talk, but to work. But, I also think that the contact with each other is very important… (woman, external)

Due to their pathology, to reach the last level, growth needs, is difficult. It is true; some members would like to assume more responsibilities, but they continue to have less autonomy. They need large amounts of support from the instructor to increase their potential.

Also, we noticed that some group members did not feel fulfilled in their work, in spite of the large number of workers that were motivated in the working place. They would like to work with other people:

> I would like to work with old people, handicapped people, and children …I would feel more fulfilled. (woman, external)

or to do housework:

> … cleaning, ironing, making beds, cooking… (woman, external)

As mentioned previously, Hackman and *Oldham* suggest a list of characteristics of the work. We are going to analyse the five dimensions and their applications in this group.

- *Skill variety:* the interviews were important in finding out the different tasks that are executed in gardening and to identify the preferred functions of each component, so the workers could do these tasks frequently. The worst task, according to the workers, is to pick weeds; it revealed that everyone has to do this kind of work. We found that some like to complete tasks, which require more physical effort. We have mentioned which persons like this kind of work. We have also suggested to the organisation to maintain the small groups of three people. The mini-group can work in a rotating way; in this case the units execute different tasks. It is important to take special care in forming the groups to avoid having people work together who have had a recent conflict. In general, these workers like to work in a group. If we left the mental patients to work alone or create groups with just two people, it would affect them psychologically:

> It begins to decrease the dialogue and increase the loneliness, isolation. For this reason, a group of three is ideal. (man, external)

With the people that prefer to work alone, we recommend that they do their work away from the mini-groups.
It would be useful if there were a daily plan, which is structured and shows the specific tasks of each worker.
- *Task identity:* at the end of each task, for example, to cut carnations, to pick weeds, etc., the workers can verify the result.
- *Task significance:* the work of the mental patients is very important in improving their self-esteem; they feel useful, it gives them more sense of meaning in their lives.
- *Autonomy:* due to their pathology, the autonomy of the mental patients is limited:

> Now I think I've changed: I think perhaps it's due to the pathology. I feel a certain limitation. Sometimes I also forget some things. (woman, external)

We have presented the elements that can carry more responsibility and autonomy. The concepts of responsibility and autonomy are connected; if a worker has more autonomy, she/he also assumes more responsibility. What we have suggested was that there will be a superior in the group when the instructor is absent. We have proposed two people that have the ability to lead. They will assume the job as chief alternately, so both can experience the difficulties that a leader can face.
- *Information:* the instructor should communicate daily about the quality of their work. This feedback is important in improving the execution of their tasks.

In regard to the expectancy theory, for Lawler the salary contributes in elevating the motivation. First, we have identified the system of payment in this organisation. It behoves us here to make a brief distinction between the internal and the external people of the Health Centre.

The internal workers receive 70 % less than the rest of their co-workers. The reason is that the external people have more expenses to support themselves. Unfortunately, this fact causes conflicts in the organisation. For example, one internal worker thinks that it is "exploitation". Another worker said:

> The internals earn less money, this causes conflicts with the externals but they (internals) have the medication of the Health Centre but they don't see that. (woman, external)

We have suggested developing a "campaign of sensibility" for the internal people because they demonstrate insufficient understanding in the differences of the wages.

Other considerations need to be mentioned. It would be useful if these workers would have sessions about motivation in order to increase their enthusiasm. For example, to discuss with the workers what can make them more motivated in their working place. We have also suggested having sessions about how to resolve conflicts between the co-workers.

Provided that we have analysed the motivation of mental patients who work in a group, we thought that it was essential to dedicate some attention to the following question:

"Does a group influence the individual or does the individual influence the group?" [3, p. 130].

We determined that the group influences the individuals. The pressure of the group makes the worker adjust his thinking and acting. The individuals are influenced at the psychological level (for example: they learn how to work in a group) and physical level:

> …the crises didn't disappear but were very sporadic… (man, external)

The individuals also influence the group:

> When the week doesn't go well, sometimes I feel that it affects my colleagues psychologically… (man, external)

Participation within a group implies the accepting of the group's rules that influence the individual's behaviour.

> I try to do it (the work) like the instructor says. (woman, external)

5 Conclusions

This study was very significant to researchers because they developed a special empathy and value for the mental patients.

The first objective of this paper was to identify the motivational degree of the mental patients who worked in gardening. For this reason, we started to interview each member of the group and fill out the inquiries. Some difficulties in using these methods were faced. Sometimes, the interviewees answered with extremely long responses or, on the contrary, some mental patients were very shy. Researchers also formulated new questions depending of the context of the interview or changed the order of the questions. Few interviewees had problems in understanding the questions. Only a small number of contradictions were found between the interviews and the inquiries.

After analysing these methods, it was concluded that most of the mental patients were motivated to do their work.

The second objective focused on making suggestions to elevate the motivation of each worker by applying motivational theories.

The principal conclusions were:

- A majority of the mental patients demonstrated characteristics of Theory Y of McGregor: the work was very significant for their well-being and was like a therapy.
- All workers, except two, were in the second level of the ERG theory of Alderfer. This means they tried to satisfy their related needs. The communication in the group was very important and the group can be a support for the mental patients.

- In regard to the model of Hackman and Oldham, we started to identify the tasks that each worker preferred, if they liked or didn't like working in a group. The recommendation is that they continue to work in a group of three, which rotated jobs. Besides, there were presented the elements of the group who can assume more responsibility and autonomy.
  Furthermore, the instructor should communicate daily as to whether the workers do their work with quality.
- In the expectancy theory of Lawler, it was identified the system of remuneration of this organisation. The internal mental patients receive less than the others. This is the reason there are conflicts between them. So, it was suggested elaborating a "campaign of sensibility" for the internal co-workers.

Other considerations that were mentioned in this research were that we thought, it would be valuable to have sessions where the workers could be motivated and learn how to resolve conflicts.

As we analysed the motivation of mental patients in a group, it was determined whether the group influences the individual or vice versa.

It was possible to verify that if, on the one hand, the group influences the co-workers in their psychological and physical level, on the other hand, the individual influences the group in the psychological level as well.

Finally, it is important to enhance that this study has given to the researchers a better understanding about the difficulties mental patients face in their work, in particular that the productivity varies according to their physical and mental health. It demands much understanding by their employers.

It will be of great relevance the managerial world to pay more attention to these special people and to be careful not to discriminate against them since they can accomplish their work well with the appropriate medications.

Bibliography

1. Ménéchal, J. (1999). *Introdução à Psicopatologia* (1st ed.). Lisboa: Psicologia de Bolso.
2. CCHR International (2014). Mental disorders: The facts behind the marketing campaign, CCHR International, Te Mental Health Watchdog, www.cchrint.org/psychiatric-disorders/. Accessed in 26 Feb 2014.
3. Rosa, L. (1994). *Cultura Empresarial, Motivação e Liderança* (1st ed.). Lisboa: Psicologia das Organizações.
4. Neves, A. L. (1998). *A Motivação para o Trabalho* (1st ed.). Lisboa: R. H. Publishing.
5. Latham, G., & Pinder, C. (2005). Work motivation theory and research at the down of the twenty-first century. *Annual Review of Psychology, 56*, 485–516.
6. Ellemers, N., De Gilder, D., & Haslam, S. A. (2004). Motivating individuals and groups at work: A social identity perspective on leadership and group performance. *Academy of Management Review, 29*(3), 459–478.
7. Caulton, J. R. (2012). The development and use of the theory of erg: A literature review. *Emerging Leadership Journeys* (Regent University School of Global Leadership & Enterpreneurship), *5*(1), 2–8.
8. Lévy-Leboyer, C. (1998). *La Motivation dans l' Entreprise, Modèles et Stratégies* (2nd ed.). Paris: Éditions d' Organisation.

9. Hackman, J. R. (1976). Motivation through the design of work: Test of a theory. *Organizational Behavior and Human Performance, 16*(2), 250–279.
10. Giancola, F. L. (2011). Examining the job itself as a source of employee motivation. *Compensation Benefits Review, 43*(1), 23–29.
11. Lawler, E. E, I. I. I. (1973). Expectancy theory and job behavior. *Organizational Behavior and Human Performance, 9*(3), 482–503.
12. Ambrose, M. L., & Kulik, C. T. (1999). Old friends, new faces: Motivation research in the 1990s. *Journal of Management, 25*(3), 231–292.
13. Steel, P., & Konig, C. J. (2006). Integrating theories of motivation. *Academy of Management Review, 31*(4), 889–913.
14. Chiavenato, I. (1991). *Recursos Humanos na Empresa, Pessoas, Organizações, e Sistemas* (2nd ed.). S. Paulo: Atlas Publishing S. A.
15. Chiavenato, I. (1977). *Introdução à Teoria Geral da Administração*. Brasil: McGraw-Hill Publishing.
16. Hackman, J. R. et al. (1980). *Work design.* Addison-Wesley Publishing Co., Inc., Boston.

Index

C. Machado and J. P. Davim (eds.), *Work Organization and Human Resource Management*, Management and Industrial Engineering,
DOI: 10.1007/978-3-319-06376-8, © Springer International Publishing Switzerland 2014

MIX
Papier aus verantwortungsvollen Quellen
Paper from responsible sources
FSC® C105338
FSC
www.fsc.org

If you have any concerns about our products,
you can contact us on
ProductSafety@springernature.com

In case Publisher is established outside the EU,
the EU authorized representative is:
Springer Nature Customer Service Center GmbH
Europaplatz 3, 69115 Heidelberg, Germany

Printed by Libri Plureos GmbH
in Hamburg, Germany